Intelligent Systems Reference Library

Volume 185

Series Editors

Janusz Kacprzyk, Polish Academy of Sciences, Warsaw, Poland

Lakhmi C. Jain, Faculty of Engineering and Information Technology, Centre for Artificial Intelligence, University of Technology, Sydney, NSW, Australia, KES International, Shoreham-by-Sea, UK; Liverpool Hope University, Liverpool, UK

The aim of this series is to publish a Reference Library, including novel advances and developments in all aspects of Intelligent Systems in an easily accessible and well structured form. The series includes reference works, handbooks, compendia, textbooks, well-structured monographs, dictionaries, and encyclopedias. It contains well integrated knowledge and current information in the field of Intelligent Systems. The series covers the theory, applications, and design methods of Intelligent Systems. Virtually all disciplines such as engineering, computer science, avionics, business, e-commerce, environment, healthcare, physics and life science are included. The list of topics spans all the areas of modern intelligent systems such as: Ambient intelligence, Computational intelligence, Social intelligence, Computational neuroscience, Artificial life, Virtual society, Cognitive systems, DNA and immunity-based systems, e-Learning and teaching, Human-centred computing and Machine ethics, Intelligent control, Intelligent data analysis, Knowledge-based paradigms, Knowledge management, Intelligent agents, Intelligent decision making, Intelligent network security, Interactive entertainment, Learning paradigms, Recommender systems, Robotics and Mechatronics including human-machine teaming, Self-organizing and adaptive systems, Soft computing including Neural systems, Fuzzy systems, Evolutionary computing and the Fusion of these paradigms, Perception and Vision, Web intelligence and Multimedia.

** Indexing: The books of this series are submitted to ISI Web of Science, SCOPUS, DBLP and Springerlink.

More information about this series at http://www.springer.com/series/8578

Jagannath Singh · Saurabh Bilgaiyan ·
Bhabani Shankar Prasad Mishra ·
Satchidananda Dehuri
Editors

A Journey Towards Bio-inspired Techniques in Software Engineering

 Springer

Editors
Jagannath Singh
School of Computer Engineering
KIIT University
Bhubaneswar, Odisha, India

Saurabh Bilgaiyan
School of Computer Engineering
KIIT University
Bhubaneswar, Odisha, India

Bhabani Shankar Prasad Mishra
School of Computer Engineering
KIIT University
Bhubaneswar, Odisha, India

Satchidananda Dehuri
Department of Information
and Communication Technology
Fakir Mohan University
Balasore, Odisha, India

ISSN 1868-4394   ISSN 1868-4408 (electronic)
Intelligent Systems Reference Library
ISBN 978-3-030-40930-2   ISBN 978-3-030-40928-9 (eBook)
https://doi.org/10.1007/978-3-030-40928-9

This Springer imprint is published by the registered company Springer Nature Switzerland AG
The registered company address is: Gewerbestrasse 11, 6330 Cham, Switzerland

Preface

Since, from its invention, Software Engineering (SE) undergoes many phases of changes and improvement. Starting from procedural software to object-oriented software, the changes occurred make SE more strengthen. The journey of SE not stopped there. It is still continuing with new paradigms such as Aspect-Oriented software and Component-Oriented software development. It was observed that, as the software development technique is improving, the other related activities such as testing, debugging, estimation and maintenance become more complex and difficult. Hence, as the software development paradigm is evolving, there is a need of regular updates of research work to ease the transition process from one paradigm to another. With the huge traces of traditional software, most of the companies have developed a repository of previously developed software and their details. Now, the platform is set for new engineering techniques such as bio-inspired optimization and searching techniques, to be introduced into each field of software engineering. Starting from the analysis of gathered customer requirement to maintenance of the software, researchers have started applying the genetic algorithm, fuzzy graph theory, artificial neural networks technique, etc. In order to make the realization become true, this volume entitled a journey towards bio-inspired techniques in software engineering has been taken into shape with an inclusion of 10 chapters contributed by potential authors.

In Chap. 1, the author focuses on transformed models of Closeness Factor Based Algorithm (CFBA), its applications in real-time systems. Software development life cycle (SDLC) and software model for Distributed Incremental Closeness Factor Based Algorithm (DICFBA) variants are also discussed in this chapter. In Chap. 2, the author discusses the design and implementation of a new logic introduced in on-board software, to overcome Liquid Apogee Motor (LAM) termination, in case of failure of sensor data updates. It also highlights the recovery time for various combinations of sensor failures.

In Chap. 3 authors developed a MATLAB program that handles the behaviour of Differential-Drive Pioneer P3-DX Wheeled Robot (DDPWR) in the V-REP software engineering platform. They have also done a comparison study between proposed Type-2 Fuzzy Controller (T2FC) technique with the previously developed Type-1 Fuzzy Controller to show the authenticity and robustness of the developed T2FC.

For evaluating the amount of data sharing between the methods of software module, a novel cohesion metric is proposed by the authors in Chap. 4. Authors of Chap. 5 have discussed different architectural patterns available for engineering micro services. They have also discussed on different challenges encountered while large scale applications are encountered as micro services. In Chap. 5, authors have also pointed out different tools that can be leveraged for the development, deployment, discovery and management of micro services.

In Chap. 6 authors have proposed a Chaos-based Modified Morphological Genetic Algorithm for Effort Estimation in Agile Software Development. In Chap. 7, authors have proposed and tested a frame work using machine learning classification technique for malware detection. Effort estimation is one of the important tasks in the field of software engineering. In Chap. 8, authors have proposed a soft computing based approach for calculation of the effort. Again a Case study on NASA93 and Desharnais Datasets has also been done in this chapter. Test data generated through path testing can also be exercised for mutation analysis. A Genetic Algorithm based approach, Test Data Generation and Optimization for White Box Testing, is discussed by the authors in Chaps. 9 and 10 authors proposed a machine learning based frame work for detection of web service anti patterns.

Bhubaneswar, Odisha, India   Jagannath Singh
Bhubaneswar, Odisha, India   Saurabh Bilgaiyan
Bhubaneswar, Odisha, India   Bhabani Shankar Prasad Mishra
Balasore, Odisha, India   Satchidananda Dehuri

Contents

Chapter 1
SMDICFBA: Software Model for Distributed Incremental Closeness Factor Based Algorithms

Rahul Raghvendra Joshi, Preeti Mulay and Archana Chaudhari

Abstract The number of users utilizing internet services per day is in billions today. Also with the invent of "Internet of Everything (IoE)" and "Internet of People (IoP)", the gigantic data is getting generated in real time every moment. To effectually handle, control, guide and utilize such vast amount of data in real time, it is essential to have distributed systems at place. Such distributed system for data management needs to be iterative in nature and parameter-free, so as to achieve quality decision making with prediction and or forecasting. "Distributed Incremental Closeness Factor Based Algorithm (DICFBA)" is primarily designed to accommodate ever growing data in numeric as well as text form. Assorted versions of CFBA are developed as per the need of the analysis till date. The primary purpose of all these CFBA driven incremental clustering models was to learn incrementally about embedded patterns from the given raw datasets. This research covers transformed CFBA models, its real-time varied domain applications, and future extensions for incremental classification point of view. Software development life cycle (SDLC) and software model for DICFBA (SMDICFBA) variants are also discussed in this chapter.

Keywords SDLC · Software models · Incremental clustering · DICFBA · SMDICFBA

R. R. Joshi (✉) · P. Mulay · A. Chaudhari
Symbiosis Institute of Technology (SIT), Symbiosis International (Deemed University),
Pune, India
e-mail: rahulj@sitpune.edu.in

P. Mulay
e-mail: preeti.mulay@sitpune.edu.in

A. Chaudhari
e-mail: archana.chaudhari@sitpune.edu.in

© Springer Nature Switzerland AG 2020
J. Singh et al. (eds.), *A Journey Towards Bio-inspired Techniques in Software Engineering*,
Intelligent Systems Reference Library 185,
https://doi.org/10.1007/978-3-030-40928-9_1

1

1.1 Introduction

Innovations in software engineering increases complexity of developed software products and in turn compromises the quality. By improvising efficiency of involved processes, software industries are able to focus on innovations and can reduce time to market. Developed software's can be used for [1]:

1. Improving the cause for which it is developed and to identify gaps
2. Workflows enabling decision support
3. Empower users and increase their self management capabilities
4. Engage and prepare for further productive developments.

So, developing an Information System (IS) for a specific problem is a critical task due to variety of reasons. For example mistakes in software development can have adverse effects leading to financial costs also. In this chapter, software models for distributed variants of incremental closeness factor based clustering algorithm (DICFBA) are discussed in Sect. 1.3. There would be definitely differences among the methodology adopted an academic researcher and by the software industry professional who will be developing such type of an IS. There are top twelve software development methodologies shown in the Table 1.1. In literature review, pros and cons about these software development methodologies are discussed. As CFBA is incremental in nature so, incremental and iterative software model, DevOps model for its distributed developments are also discussed in Sect. 1.3 of this chapter.

Waterfall model is meant for the novice developer [1]. Prototype model gives complete functionality feel to the client [1]. Agile development is useful when there are frequent alterations in the development project [1]. Rapid development reduces risk and efforts in the development [1]. Dynamic development is based on rapid development; it is iterative and incremental with continuous user involvement [1]. Spiral model primarily focuses on early identification and reduction of involved risks [1]. Extreme programming is basically an agile methodology intended to create software's in unstable environment [1].Each of these software development methodologies are superior in one or other aspects in comparison to other methodologies [1]. None of the methodology is full proof. Feature driven is an iterative model using object oriented technology [1]. Joint application development is carried through intense off-site meetings of end-user, clients and software developers to finalize the system [1]. Lean development produces easily changeable software's [1]. Rational

Table 1.1 Top 12 software development methodologies [1]

Waterfall model	Prototype model	Agile software development
Rapid application development	Dynamic systems development model	Spiral model
Extreme programming	Feature driven development	Joint application development
Lean development	Rational unified process	Scrum development

Table 1.2 SE tasks and applicable ML types [2]

SE tasks	Applicable type(s) of ML
Requirements analysis	Analytical learning, inductive logic programming
Prototype development	Unsupervised learning
Reuse of components	Instance based learning (case based reasoning)
Budget/manpower prediction	Instance based learning (case based reasoning), supervised learning
Defect finding	Supervised learning
Output checking	Analytical learning, explanation based learning
Validation	Analytical learning
Reverse engineering	Concept learning

unified process is the modern software development method with combo of object-oriented and web-enabled program development methodology [1]. Scrum can be nearly applied to any project [1]. Agile and Scrum are two most common development methods used by software industries due to their adaptive nature, involved feedback mechanism and lightly controlled methodologies [1]. Pros and cons of Agile, Scrum, and other methods are discussed briefly in the next section. Software engineering automates tasks by writing rules; on the other hand machine learning (ML) goes a step ahead: it automates the rule writing task. ML can be utilized in:

1. for discovering implicit valuable regularities from large database,
2. knowledge needed to develop effectual algorithms
3. and for domains that must dynamically adapt to changing conditions.

Major ML types include: analytical learning, conceptual learning, instance based learning, inductive logic programming, supervised learning, unsupervised learning and reinforcement learning. So, not surprisingly, ML models and methodologies can be used for better software developments. One of the attractive features of ML is they bid an important complement to the available gamut of tools to make it easier to go up to confront of the aforesaid challenging situations. Amalgamation of software engineering (SE) tasks and ML types can be seen in the below motioned Table 1.2.

Figure 1.1 shows clear-cut distinction between classical and ML based software developments. ML oriented software development considers data cleansing; feature selection, model selection and training followed by validation. These steps make ML oriented software development process more superior as compared to classical process. Figure 1.2 shows ML aided software product development. It is the bit elaborated version of Fig. 1.1. Figure 1.3 shows ML adoption due to involved extensive analysis and insights. The extensive analysis and insights carries more percentages as compared to other processes. Figure 1.4 shows in 2018 ML/AI based software development initiatives are at the top as compared to other initiatives. 60% of software development in 2018 is ML/AI based. Figure 1.5 shows there is 34% hike in terms ML application development during last five years i.e. from 2013 to 2017.

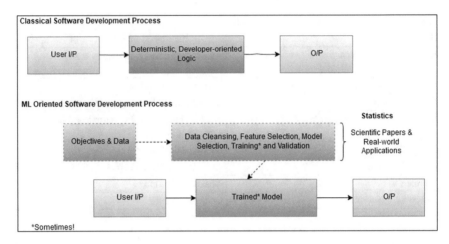

Fig. 1.1 Classical versus ML based software development [3]

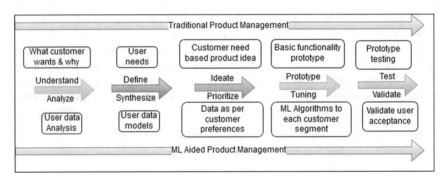

Fig. 1.2 ML aided software development [4]

Statistics in Figs. 1.4 and 1.5 are triggers behind hike in ML application development. Figure 1.6 projects that ML products development will be double up to 2020. Figure 1.6 considers case study of Deloitte Company to show doubled ML pilot projects and implementations.

So, this in a way shows strong linkage between ML and recent software development trends. The reason behind looking at details about ML types applicable to specific SE tasks, ML driven software development, tangible benefits of ML based software development in terms of analysis, insights and considerable growth in ML application development by top industries is that likewise industry in academia also now-a-day's ML drives most of the software developments by considering its virtues like data cleansing, feature selection, model selection, training and testing of data, validation of input and output.

Closeness Factor Based Algorithm (CFBA) is one-of-its kind of pure incremental data clustering algorithms [6]. In recent years, incremental learning, software engineering, data mining, incremental clustering etc. are the key research areas and have

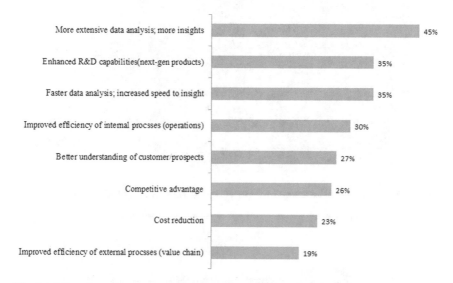

Fig. 1.3 ML-software development gives more data analysis and insights [5]

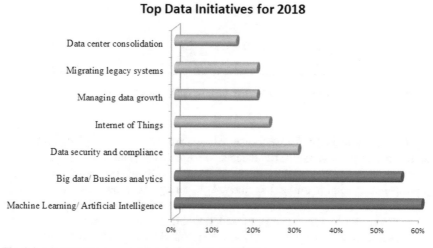

Fig. 1.4 ML/AI is the top most data initiatives in 2018 [5]

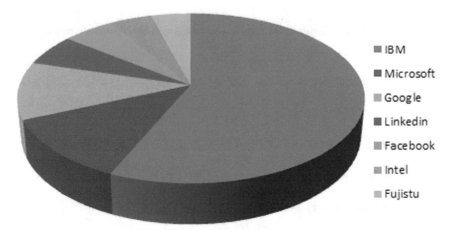

Fig. 1.5 34% growth in ML related developments by top industries [2013–2017] [5]

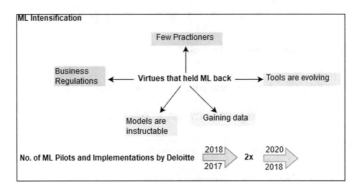

Fig. 1.6 Deloitte global predicts number ML projects will double in 2018 and 2020 [5]

received great attention. Conventional ML approaches don't consider systemic interactions; they fail in multifaceted decision scenarios. Dynamic learning scenarios, its exploration and elegant decision making can be possible through CFBA. CFBA possess incremental learning, incremental clustering and knowledge augmentation characteristics [6, 7]. This chapter covers software development life cycle (SDLC) and software models for already existed, ongoing distributed developments of incremental closeness factor based algorithms (DICFBA) on platforms like Microsoft Azure, Hadoop and Amazon Web Services (AWS). Microsoft Azure had given of 12K$ grant under data science and AI for earth themes respectively for ML oriented DICFBA specific proposals [8–10, 33]. This grant is in the form of Azure cloud subscription for one year. The Azure subscription grant was utilized to carry out deployment of DICFBA on Azure private cloud. The 12K$ Azure subscription consists of pay as you go facility for accessing Azure storage, virtual machines from different counties, ML deployment support etc. Normally 4+1 model is followed for software developments. This chapter proposes 4+1 expanded and 5+1 models for DICFBA. Recent distributed developments of CFBA focuses on disease analysis and on smart data analysis. Disease management and financial fitness both are related entities. DICFBA and financial fitness enablement are also covered in this chapter.

This book chapter will be an eye opener for researchers who want to get insights about SDLC and software models about distributed incremental clustering/software development. It will also serve as a guideline for distributed incremental clustering/software developments.

1.2 Literature Review

Software development in general involves six stages: analysis, design, implementation, testing, maintenance and documentation of developed software. As a discipline software engineering was known for more than three and half decades ago, there had been variety of definitions of software engineering [11] in the literature. "The form of engineering that uses computer science principles to achieve cost efficient software solutions" definition by Software Engineering Institute (SEI) at Carnegie Mellon University (CMU) [12]. A systematic, disciplined, quantifiable approach applied to the development, operation, and maintenance of software-definition by IEEE computer society and by ACM for software engineering [13]. Software engineering in general considers a group of tools, techniques, describing each development phase, not specifying details. These details can be tailored as per needs of an organization. The software engineering framework can also be used much more at abstract level for example for design or for specifying product configuration.

Table 1.1 shows top 12 software development practices. Pros and cons of few common software developments practices need to be discussed. They are as follows:

1. Agile model is set of principles and practices that minimize documentation, intended to do activities with high professionalism and the development time factor is the main advantage [14]. High knowledge and experienced team are the drawbacks of agile model [14].
2. On the other hand scrum is lightweight, simple to understand [15]. Scrum does not bring out excellence; it exposes incompetence [15].
3. Another commonly followed software development model is waterfall in which sequential design is followed using six development stages [16]. It is time-consuming and unclear requirements at the early stages are the drawbacks, easy to model and implement are the advantages of waterfall model [16].
4. Iterative model is the simple execution of a subset of the software needs and iteratively enhances the developing versions until the full system is implemented [7]. Modifications to design are made; new serviceable capabilities are incorporated into each iterations [17].

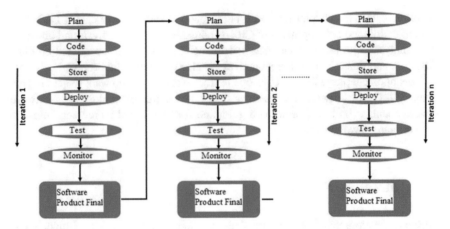

Fig. 1.7 Six stages iterative incremental software development model inspired by DevOps for distributed clustering algorithms

ICFBA follows both iterative cycles and incremental updates. So, Iterative and Incremental which is combo of both iterative design and incremental model is considered for ICFBA developments.

Strengths of iterative incremental software development model are: prioritization of requirements, fast delivery, and important functionality can be developed early, lowering delivery cost, release is product increment specific, customer feedback after each increment and easy accommodation of requirements is possible. Weakness of iterative incremental model are: effectual planning of iterations, efficient design enabling provision for inclusion of changes later, early definition of a complete fully functional system and well defined module interfaces allowing definition of increments need to be done carefully.

Now-a-day's DevOps is mainly and largely used for distributed software developments [18]. Figure 1.7 shows six stages iterative incremental software development model inspired by DevOps for distributed clustering algorithm. In Sect. 1.3, mind map for six stages DevOps SDLC which can be mapped to DICFBA is shown (Fig. 1.7).

As the main thrust of this book chapter is software models for DICFBA, so there is need for prologue about CFBA/ICFBA. Data is continuously generated through variety of applications viz. through monetary transactions; flow in energy networks, satellite imagery and through web processing. To handle such a voluminous data with data increments at every second, there is a need of incremental clustering approach. Such incremental clustering approaches leads to incremental learning and augmentation of knowledge at different stages.

Table 1.3 Till date CFBA developments

Sr. no.	CFBA developments	Year	Datasets used
1	Enhanced closeness factor algorithm for effectual forecasting (ECFAEF) [19]	2010	Organizational database
2	Probabilistic CFBA [20]	2013	Wine, software development life cycle
3	Modified cluster formation algorithm (MCFA) [21]	2015	Ice cream, diabetes patients datasets
4	Incremental clustering Naïve bayes closeness-factor algorithm (ICNBCFA) [22]	2016	Wine, electricity, software project, zenni optics and wine quality
5	Correlation-based incremental clustering algorithm (CBICA) [23]	2017	Wine
6	Threshold based clustering algorithm (TBCA) [24]	2017	Diabetes mellitus (DM) patients dataset
7	Nature inspired (version of CFBA) ants feeding birds algorithm [25]	2017	Document clustering
8	Deep incremental statistical closeness factor-based algorithm (DIS-CFBA) [26]	2018	DM dataset
9	Cloud4ICFBA: Microsoft Azure based distributed incremental closeness factor based clustering algorithm (DICFBA) [10]	2019	DM dataset

1.2.1 Closeness Factor Based Algorithm (CFBA)

CFBA is the prominent candidate for tackling such incremental information update during such scenarios. CFBA came into existence in 2010, after that it gets evolved into different versions as per specific needs. Table 1.3 enumerates till date developments of CFBA/ICFBA.

If two data points are close to each other then they are likely belong to the same cluster-CFBA/ICFBA works on this ideology. Clustering algorithms are of two types distance based or pattern based. CFBA is a pattern based clustering algorithm that's why it does not suffer from dimensionality issues. CFBA handles both structured and unstructured data types and being domain independent. CFBA is a parameter-free algorithm—no need of initialization condition or selecting an initial value of K etc. CFBA during its processing considers existing data first and on arrival of new data, old data either get updated or new clusters are formed. This in-a-way is useful for updating old learning, building new one and does knowledge augmentation. CFBA operated in semi supervised and selective incremental processing mode. Correlation Based Incremental Clustering Algorithm (CBICA) is the another variant of CFBA where instead of probability as in case of CFBA, Pearson correlation value is used to calculate the CF. Another variants of CFBA are based on Naïve Bayes, Threshold, Statistical and in case of MCFA distance measures are used though CFBA is free from distance measures.

After looking at prologue and till date developments about CFBA/ICFBA, it is necessary to look at transformation of CFBA from singleton to distributed mode. Apart from CFBA developments specified in Table 1.3. Recently, DICFBA is also proposed in the form of internet of medical things (IoMT). Next section consolidates single machine, distributed platform specific viz., Hadoop, AWS, Microsoft Azure, IoMT specific software models for ICFBA. As discussed in this section and it is evident from Table 1.3 also ICFBA supports cloud specific developments also. DevOps SDLC for DICFBA is also discussed in Sect. 1.3. Also, different scenarios from IoMT and distributed developments point view are also covered in upcoming section. So, in other words next section throws light on journey of CFBA towards its till distributed development from software engineering models perspective. As discussed at the beginning of this chapter there a linkage between CFBA/ICFBA and financial fitness which is also elaborated as a model in continuing section.

1.3   Proposed Software Models for Distributed Incremental Closeness Factor Based Algorithms (SMDICFBAs)

Before going to discuss SMDICFBA's, let us walk through singleton CFBA software model. As shown in Fig. 1.8, singleton CFBA does processing through following steps:

1. Cleansing of raw dataset
2. Impactful attributes identification through Principle Component Analysis (PCA)/ Independent Component Analysis (ICA) or Sparse PCA (SPCA)
3. Based on influx of data, update old clusters or new clusters are formed
4. In post clustering phase by analyzing formed clusters validation of impactful attributes at the beginning and at the end of CFBA processing are done
5. Three phase working—"process-append-modify"
6. Iterative and incremental software model is followed by singleton CFBA.

Singleton CFBA software developments follow 4+1 software architecture model as shown in Fig. 1.10. Process view for CFBA is involved operations that calculate closeness factor (CF) between two data series. CF can be either 0 or 1. 1 means alike and 0 means data series are dissimilar to each other. Physical view is basically console that prompts CF calculations. Development view is basically tools and technologies used for CFBA execution. Logical view is the UI that facilitates user about impactful attributes validation, visualization of formed clusters, accuracy details after specified iterations etc for CFBA. So, it is evident that software project management and end user functionalities can be achieved through 4+1 software architecture model for CFBA for a particular development.

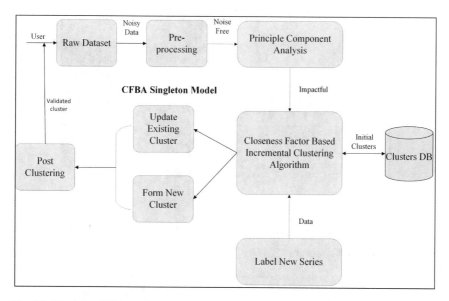

Fig. 1.8 Singleton CFBA model

One more variant of singleton CFBA is there i.e. CBICA. CBICA gives correlation value among two data series which is either −1/1 or 0 (refer Fig. 1.9). Pearson correlation is used in CBICA. In order to make it CFBA probability independent, CBICA is invented. The correlation value is related to the closeness factor. Closeness value ranges from 0 to 1 to form clusters, to solve this issue by Pearson's coefficient of correlation similarity measures which ranges from −1 to 1. So that negatively correlated data is also processed by variant of CFBA i.e. CBICA. Likewise CFBA, CBICA also follows 4+1 software architecture model for considered development. One more added advantage of CBICA is that negative relation among data series can also be analyzed. Almost in all cases, positive CF values are exhibited by CFBA, except in case of ionosphere dataset from UCI repository [27]. These CF or correlation values can act as a relevance check or feedback loop for CFBA/ICFBA or CBICA/ICBICA specific software developments. Apart from CFBA/CBICA, Threshold based clustering algorithm is the third variant of singleton CFBA. Based on cluster average versus outlier average, cluster deviation versus outlier deviation, TBCA gives impactful attributes. After looking at details about single CFBA model and the software architecture followed by them. It is necessary to look at SDLC for singleton CFBA variants (Fig. 1.10).

Now, we are going look at phase specific activities that need to be performed during SDLC for singleton CFBA variants. Table 1.4 shows SDLC for singleton CFBA's variants.

Singleton implementations are not apt in today's scenario to handle large quantum of data which is continuously generated; this quantum is beyond processing powers of singleton implementation. CFBA/ICFBA exhibit in distributed form (refer

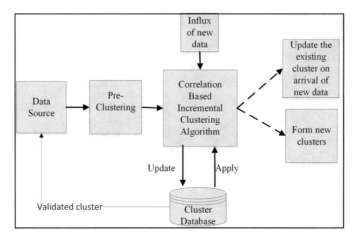

Fig. 1.9 Singleton CBICA model

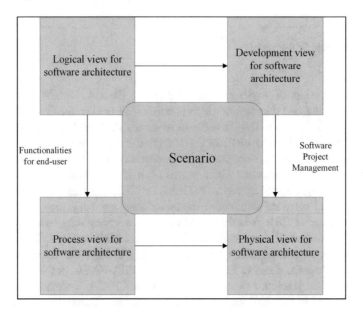

Fig. 1.10 4+1 software architecture model

Table 1.3). ICFBA is recently deployed on Microsoft Azure as Cloud4ICFBA. Also, its implementation on AWS and Hadoop are in process. As discussed in previous section now-a-day's most of the distributed software developments follow DevOps SDLC, so, DevOps SDLC for distributed variants of ICFBA is elaborated in detail in later of part of this section. Before going to discuss SDLC, software models, different scenarios for distributed ICFBA variants, it is necessary to go through the properties that make ICFBA distributable. CFBA not only support distributable characteristics

Table 1.4 SDLC for singleton CFBA's

Sr. No.	SDLC phase	Activities to be performed
1	Analysis	Selection of dataset, No. of iterations to be performed, No. of phases to be worked out
2	Design	System architecture, DFDs, process flow diagram as per implementation
3	Implementation	Finalization of tools, IDEs, technologies to be used as per selected variant
4	Testing	Cluster computation, threshold, impactful attributes and their ranges
5	Maintenance	Making specific singleton variant dataset specific, fine tuning with phased working approach
6	Documentation	Singleton variant specific algorithms, performance evaluation

but also supports distributable validations too. Apart from standard set of distributed validations, DICFBA has its own distributed validations checks like cross verification of impactful attributes at the beginning and end, cluster threshold etc. Likewise singleton CFBA variants, DICFBA variants too support structured, unstructured datasets. ICFBA's properties making it compatible for distributable implementation are as follows:

1. Full converge over input dataset.
2. Linear clustering approach-linear similarity detection between clustering, handling of linear queues.
3. Communication and synchronization dependents up on selected system configuration.
4. Frequent re-training is not required so workflow will not be interrupted in case of CFBA.
5. Load balancing, auto-scaling, scaling and fault tolerance depends upon selected framework and platform.
6. No issues about selection of coordinator and value agreement as coordinator will perform CF calculations. CF or correlation values will be exchanged between master and slave.

So, ICFBA can be made distributable on different platforms and frameworks. Mohammad Hamdaqa et al. in 2014 proposed 5+1 software architecture model for cloud applications [28] shown in Fig. 1.11. This 5+1 software architecture model is followed by cloud based DICFBA applications in the following manner:

1. Cloud based DICFBA model can be accessed, audited and evaluated against selected cloud configuration using availability meta-model
2. Azure or AWS allows to create a task group, such a task group facilitates rules for adaptation and actions for activity or task group, Azure or AWS DICFBA implementations make use of adaption meta-model while creating task groups

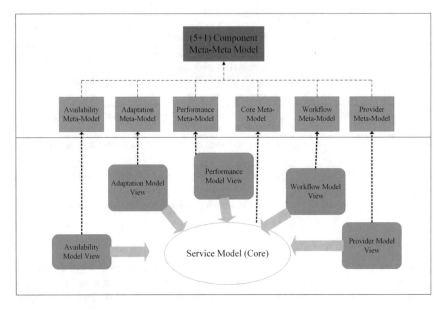

Fig. 1.11 5+1 software architecture model for cloud applications

3. Performance evaluation of cloud based DICFBA against scalability, load-balancing, scale up or scale down etc. is done using performance-metal model
4. Frontend, backend design and storage connections, storage specific interactions with UI are performed by considering core-meta model for cloud based DICFBA
5. Training of cloud based DICFBA against specific dataset is done using workflow-meta model
6. Available zones for selection of VMs, their service templates selection can be done using provider-meta model

So, this in a way validates 5+1 software architecture model also fits for cloud based DICFBA implementations.

As discussed in the introduction and in literature review sections of this chapter, apart from cloud based DICFBA implementations if anyone wants to test dataset on physically distributed ICFBA implementation like on Hadoop, 4+1 expanded software architecture model suitable to both physically distributed and cloud or web app based distributed scenarios is proposed and followed (refer Fig. 1.12). This 4+1 expanded software architecture model for generic DICFBA implementation covers all of its facets viz. process, physical, logical and developmental view in terms of intended activities for DICFBA. These days ML enabled software developments are at boom both at industry and academia level. In the literature different types of in clustering algorithms are available viz. Incremental K-Means [29], Incremental DB-SCAN [30], Incremental SubSpace clustering [31] etc. These algorithms got evolved in terms to distributed versions like distributable incremental K-Means, distributable

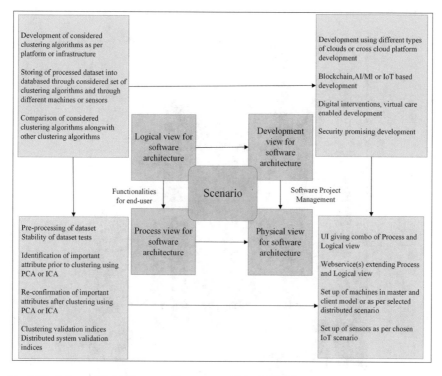

Fig. 1.12 4+1 expanded software architecture model for DICFBA

incremental DBSCAN and so on. In major cases, there exist a disconnect between singleton and distributable implementation of specific clustering algorithm. This in turn makes difficult to understand SDLC, software architecture model followed by these implementations. This chapter is one of its attempts which consolidates journey of an incremental clustering algorithm through its transformation from desktop version to distributed mode, covering software architecture model for single and distributable models as well, presenting respective SDLCs. So, this chapter will be a set of guidelines for the researchers who want to build, transform their incremental software applications to distributed one.

After looking at details about 5+1, 4+1 expanded software architecture models for DICFBA, it is necessary to look at DevOps SDLC applicable to cloud based DICFBA implementations. Four plausible scenarios for DICFBA implementations, software models from DICFBA, and DICFBA-IoMT point of view for platforms like Microsoft Azure, AWS and Hadoop are also covered. Figures 1.16, 1.17, 1.18, 1.19, 1.20, 1.21 and 1.22 depicts software models for DICFBA and DICFBA-IoMT. These figures give clear-cut idea about formulation of specific DICFBA implementations. Also, this chapter throws light on linkage between financial fitness and disease related DICFBA implementation-which would be eventually an upcoming research topic covered in later part of this section. Figures 1.13, 1.14, 1.15, 1.16 and 1.17 show different

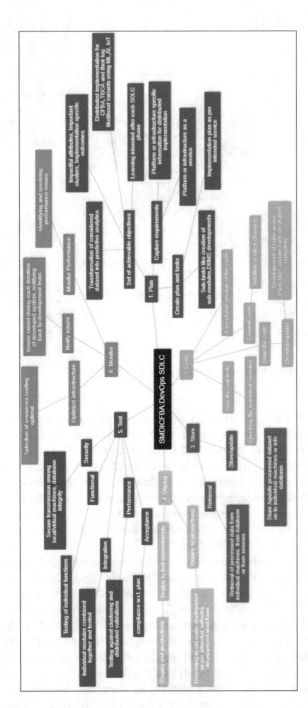

Fig. 1.13 DevOps SDLC for DICFBA

Fig. 1.14 DICFBA scenario 1: consolidated data input from multiple devices

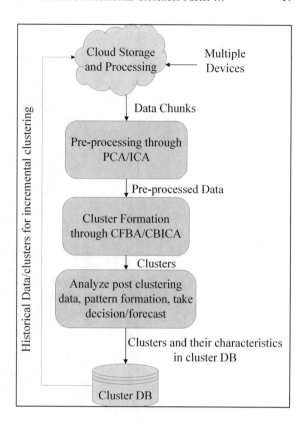

scenarios that need to be considered for DICFBA implementation. Figures 1.14 and 1.15 shows how data from multiple and individual devices can be considered as an input for DICFBA. Figure 1.16 show how data from individual devices can be given as input to cloud variants of CFBA or CBICA or cloud based DICFBA or DICBICA. On the other hand, Fig. 1.17 shows how output data from devices embedded with CFBA/CBICA can be given as output to distributed variants of CFBA/CBICA.

After looking at four different types of scenarios for DICFBA implementations, Figs. 1.18, 1.19, 1.20 and 1.21 elaborate platform specific software models for DICFBA. Figure 1.18 shows Azure based DICFBA deployment; Fig. 1.19 elaborates internal working of Azure VM. Figure 1.20 shows AWS based DICFBA implementation. Figure 1.21 shows Hadoop MapReduce based DICFBA model.

Figures 1.22, 1.23 and 1.24 elaborates IoMT specific DICFBA implementations. Likewise DICFBA, these diagrams are also platform specific. Figures 1.25, 1.26 and 1.27 shows extensibility of CFBA/ICFBA/DICFBA or CBICA/ICBICA/DICBICA for financial fitness and for document clustering etc. Figure 1.28 shows third variant of DICBICA for document clustering it is by using socio-inspired version of CFBA called as Ants Feeding Birds distributed algorithm [25].

Fig. 1.15 DICFBA scenario 2: input data from multiple devices

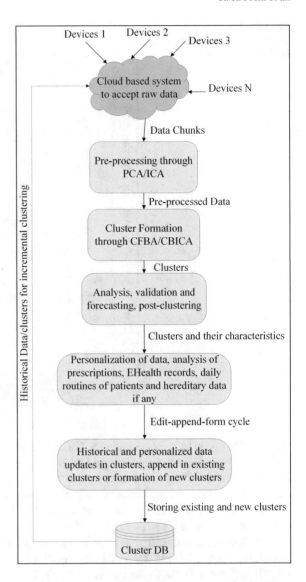

Fig. 1.16 DICFBA scenario 3: input data from individual devices supplied to cloud variants of CFBA/CBICA

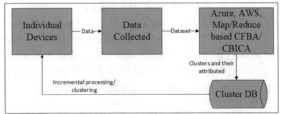

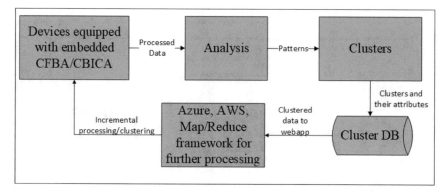

Fig. 1.17 DICFBA scenario 4: output data from devices equipped with CFBA/CBICA supplied to their distributed variants

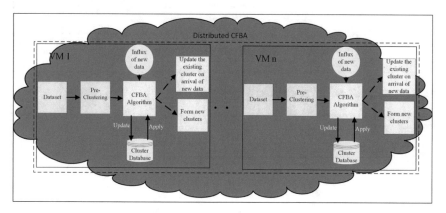

Fig. 1.18 DICFBA Azure model

1.3.1 Performance Evaluation of Proposed SMDICFBA

As discussed in introduction and literature review sections, deployment of ICFBA as DICFBA is done as Cloud4ICFBA, below mentioned Table 1.5 shows comparison of Cloud4ICFBA (one of the SMDICFBA variants) with other distributed incremental variants i.e. DI-Means (Distributed Incremental Means/K-Means) and DIDenseScan (Distributed Incremental Density Scan). 45K Diabetes mellitus (DM) pathology reports are analyzed through Cloud4ICFBA, DI-Means and DIDenseScan and it is observed that Cloud4ICFBA outperforms over other two algorithms.

The cluster which contains more number of elements is considered as priority cluster. C1/C2 is the priority cluster for DM analysis. DM dataset consists of 8 attributes Blood glucose fasting (mg/dl), Blood glucose PP (mg/dl), Cholesterol (mg/dl), Triglycerides (mg/dl), HDL cholesterol (mg/dl), VLDL cholesterol (mg/dl), LDL cholesterol (mg/dl) and NON HDL cholesterol (mg/dl). Out of these eight

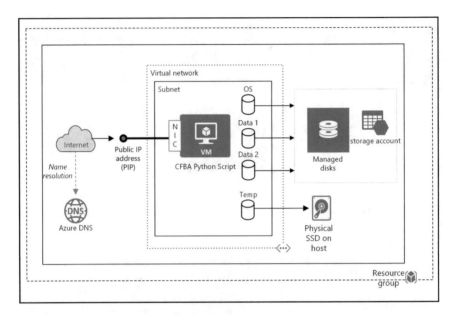

Fig. 1.19 DICFBA Azure VM's internal working [32]

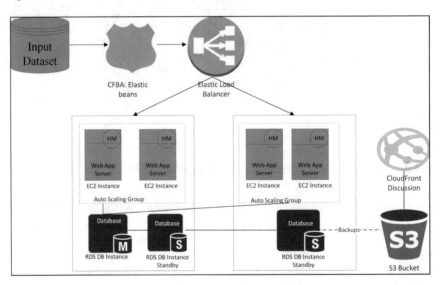

Fig. 1.20 DICFBA AWS model

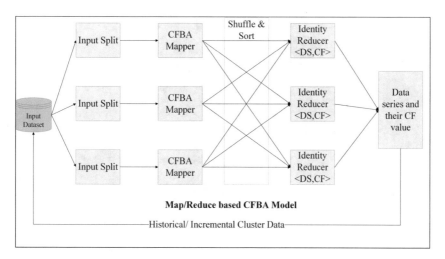

Fig. 1.21 DICFBA Hadoop model

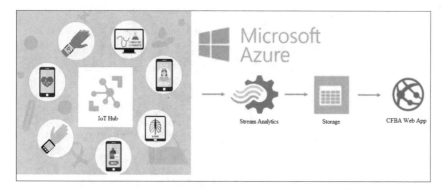

Fig. 1.22 DICFBA-IoMT Microsoft Azure model

attributes, first four attributes are impactful/principal components. Priority clusters are driven by these principle components or impact attributes. These priority clusters are also helpful in providing personalized treatment to diabetic patients.

1.4 Conclusive Summary

Conventional SDLC models are known for decades now. It is necessary to design and propose new learning models to accommodate high volumes of real time data, from varied domains. Through this book chapter we have attempted the proposal of Machine learning models, with special focus on CFBA algorithm and its variants. With this work it was proved that CFBA is distributable and a perfect applicant

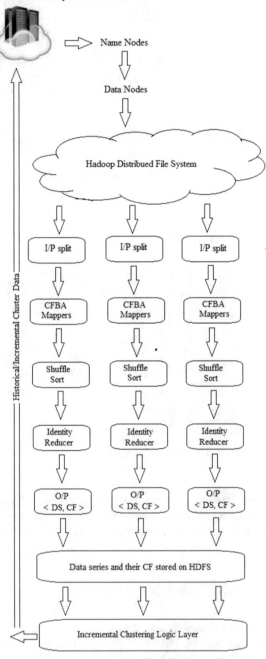

Fig. 1.23 DICFBA-IoMT Hadoop model

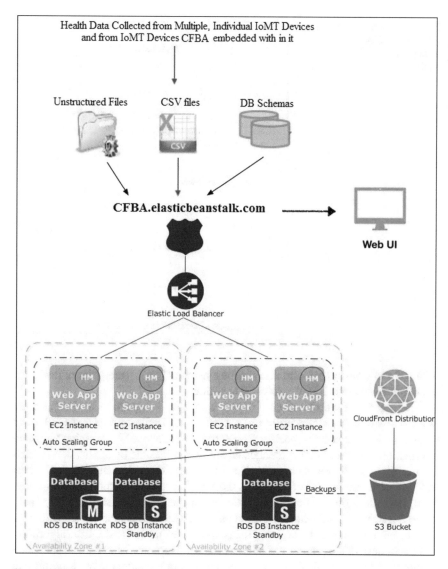

Fig. 1.24 DICFBA-IoMT AWS model

for various domains where data in the form of numbers or text is handled in large scale. FinTech is major area of vast data analysis tasks and Distributed CFBA with subspace clustering approach, its feasible to predict and provide financial fitness for an individual investor, organization and all other finance player in the market today. Personalization about individual's financial fitness can be achieved using Distributed CFBA models as proposed in this chapter. Financial fitness of an individual is directly or indirectly mapped to health by which incremental learning can be achieved about

Fig. 1.25 FinTech CFBA IS
useful for building
moneycontrol/wellness
profile

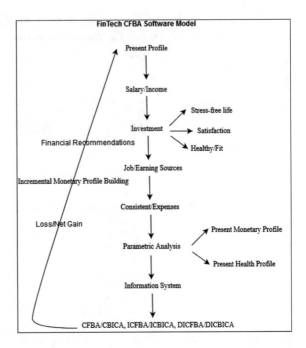

influx of new investment instruments in the market and about categories of investors, as both of them are mutually exclusive in nature.

Similarly the new text variant of CFBA is capable enough to locate misinformation from various web healthcare resources/URLs. Application of text-CFBA can be disease specific or area specific. Recommendation about URLs giving validated information and written by authenticated sources is feasible by application of text-CFBA. Text-CFBA works best with Google like search engine and also is the best fit for Clusty.com. In case of Clusty, as the search results are displayed in clustered form, meta-clustering and multi-heuristic details can be furnished effectually about the search string. Incremental learning about the category of people devoting time for misinformation also can be captured using special features of text-CFBA algorithm.

In summary, it's essential to understand the importance of new entrants called Incremental learning SDLC models, wherein learning continuously based on real time data is the major outcome.

Fig. 1.26 CFBA/ICFBA/
DICFBA based text
clustering from URLs

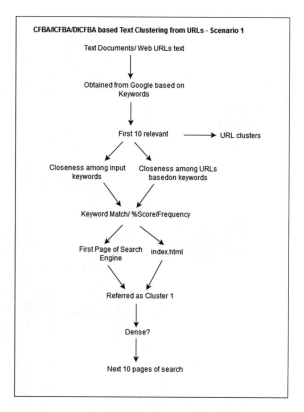

Fig. 1.27 CFBA/ICFBA/
DICFBA based text
clustering from clusty.com

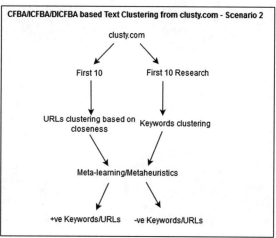

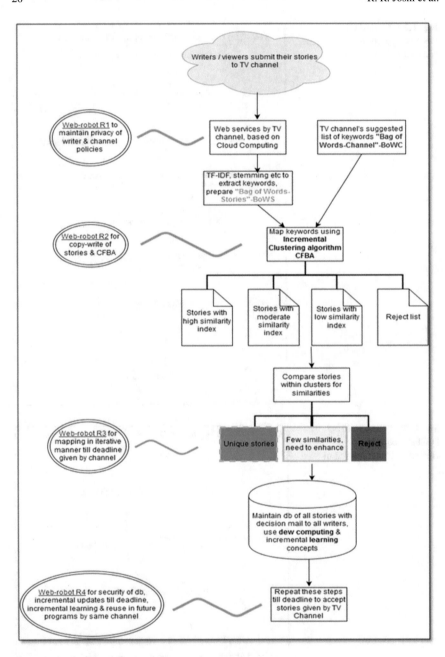

Fig. 1.28 Ants feeding birds document clustering algorithm [25]

Table 1.5 Performance evaluation of Cloud4ICFBA (SMDICFBA variant)

Input size	DM dataset	Cloud4ICFBA	DI-means (K = 4)	DIDenseScan
45K	45K	C1:24701, C2:15869, C3:4430	C1:15805, C2:12695, C3:9998, C4:6502	C1:22490, C2:9863, C3:12647
	1-22.5K	C1:10214, C2:10108, C3:2178	C1:6412, C2:5713, C3:4838, C4:5537	C1:10346, C2:6156, C3:5980
	22501-45K	C1:9826, C2:10496, C3:1537, C4:641	C1:3982, C2:7268, C3:4613, C4:6637	C1:5303, C2:6351, C3:5017, C4:5824
Average accuracy		90%	83%	80%

References

1. https://www.tatvasoft.com/blog/top-12-software-development-methodologies-and-its-advantages-disadvantages/. Accessed 5 July 2019
2. Zhang, D.: Applying machine learning algorithms in software development. In: Proceedings of the 2000 Monterey Workshop on Modeling Software System Structures in a Fastly Moving Scenario, pp. 275–291 (2000)
3. https://blog.algorithmia.com/building-intelligent-applications/. Accessed 5 July 2019
4. http://arunkottolli.blogspot.com/2018/06/how-machine-learning-aids-new-software.html. Accessed 5 July 2019
5. https://www.forbes.com/sites/louiscolumbus/2018/02/18/roundup-of-machine-learning-forecasts-and-market-estimates-2018/. Accessed 5 July 2019
6. Kulkarni, P.A., Mulay, P.: Evolve systems using incremental clustering approach. Evol. Syst. **4**(2), 71–85 (2013)
7. Joshi, R.R., Mulay, P.: Closeness factor based clustering algorithm (CFBA) and allied implementations-proposed IoMT perspective. In: A Handbook of Internet of Things in Biomedical and Cyber Physical System, pp. 191–215. Springer, Cham (2020)
8. https://news.microsoft.com/en-in/features/microsoft-ai-for-earth-grant-recipients-india/. Accessed 6 July 2019
9. Chaudhari, A., Mulay, P.: A bibliometric survey on incremental clustering algorithm for electricity smart meter data analysis. Iran J. Comput. Sci. 1–10 (2019)
10. Joshi, R.R., Mulay, P.: Cloud4ICFBA: microsoft azure based distributed incremental closeness factor based clustering algorithm (DICFBA) for analysis of diabetes mellitus
11. Al-Dahmash, A., El-Masri, S.: A new proposed software engineering methodologyfor healthcare applications development. Int. J. Mod. Eng. Res. **3**(3), 1566–1570 (2013)
12. Richardson, W.E.: Undergraduate software engineering education. In: SEI Conference on Software Engineering Education, pp. 121–144. Springer, New York (1988)
13. Board, I.S.: IEEE Standard Glossary of Software Engineering Terminology. The Institute of Electrical and Electronics Engineers, New York (1990)
14. https://www.tutorialspoint.com/sdlc/sdlc-agile-model. Accessed 6 July 2019
15. https://xbsoftware.com/blog/software-development-life-cycle-sdlc-scrum-step-step/. Accessed 6 July 2019
16. https://www.tutorialspoint.com/sdlc/sdlc-waterfall-model. Accessed 6 July 2019
17. https://www.tutorialspoint.com/sdlc/sdlc-iterative-model. Accessed 6 July 2019
18. https://aws.amazon.com/devops/what-is-devops/. Accessed 6 July 2019
19. Kulkarni, P.A., Mulay, P.: Behavioral pattern generation and analysis: application of enhanced closeness factor algorithm for effectual forecasting. SCIT J. **1**(10), 49–56 (2013)
20. Mulay, P., Kulkarni, P.A.: Knowledge augmentation via incremental clustering: new technology for effective knowledge management. Int. J. Bus. Inf. Syst. **12**(1), 68–87 (2013)

21. Gaikwad, S.M., Mulay, P., Joshi, R.R.: Mapping with the help of new proposed algorithm and modified cluster formation algorithm to recommend an ice cream to the diabetic patient based on sugar contain in it. Int. J. Stud. Res. Technol. Manag. **3**(6), 410–412 (2015)

22. Mulay, P.: Threshold computation to discover cluster structure: a new approach. Int. J. Electr. Comput. Eng. **6**(1), 275 (2016)

23. Shinde, K., Mulay, P.: CBICA: correlation based incremental clustering algorithm, a new approach. In: 2017 2nd International Conference for Convergence in Technology (I2CT), pp. 291–296. IEEE (2017)

24. Mulay, P., Joshi, R.R., Anguria, A.K., Gonsalves, A., Deepankar, D., Ghosh, D.: Threshold based clustering algorithm analyzes diabetic mellitus. In: Proceedings of the 5th International Conference on Frontiers in Intelligent Computing: Theory and Applications, pp. 27–33. Springer, Singapore (2017)

25. Mulay, P., Patel, K., Gauchia, H.G.: Distributed system implementation based on ants feeding birds algorithm- electronics transformation via animals and human. In: Detecting and Mitigating Robotic Cyber Security Risks, pp. 51–85. IGI Global (2017)

26. Joshi, R.R., Mulay, P.: Deep incremental statistical closeness factor based algorithm (DIS-CFBA) to assess diabetes mellitus. Blood **115**, 210 (2018)

27. https://archive.ics.uci.edu/ml/datasets/ionosphere. Accessed 6 July 2019

28. Hamdaqa, M., Tahvildari, L.: The (5 + 1) architectural view model for cloud applications. In: Proceedings of 24th Annual International Conference on Computer Science and Software Engineering, pp. 46–60. IBM Corporation (2014)

29. Pham, D.T., Dimov, S.S., Nguyen, C.D.: An incremental K-means algorithm. Proc. Inst. Mech. Eng. Part C: J. Mech. Eng. Sci. **218**(7), 783–795 (2004)

30. Bakr, A.M., Ghanem, N.M., Ismail, M.A.: Efficient incremental density-based algorithm for clustering large datasets. Alex. Eng. J. **54**(4), 1147–1154 (2015)

31. Jahirabadkar, S., Kulkarni, P.: ISC-Intelligent subspace clustering, a density based clustering approach for high dimensional dataset. World Acad. Sci., Eng. Technol. **55**, 69–73 (2009)

32. Run a Windows virtual machine on Azure, https://docs.microsoft.com/en-us/azure/architecture/reference-architectures/n-tier/windows-vm. Accessed 6 July 2019

33. Chaudhari, A., Joshi, R.R., Mulay, P., Kotecha, K., Kulkarni, P.: Bibliometric survey on incremental clustering algorithms. Libr. Philos. Pract., 1–23 (2019)

Chapter 2
A Novel Method for Fault Tolerance Intelligence Advisor System (FT-IAS) for Mission Critical Operations

M. Balaji, C. Mala and M. S. Siva

Abstract In Communication and inter-planetary missions, satellites are placed in elliptical parking orbits (EPO). This is followed by a series of maneuvers which subsequently positions the spacecraft in the de-sired Orbit. Liquid Apogee Motor (LAM) mode is a thruster firing mode used for orbit raising with the help of sensors such as Dynamically Tuned Gyroscopes (DTG), Digital Sun Sensors (DSS), Star Sensors (SS) and actuators such as thrusters. In the LAM mode, the output from the selected sensor is used to update the spacecraft attitude and this is compared with the desired attitude steering profile to derive the error in the attitude. The controller then corrects the spacecraft attitude error along the three axes. Presently, in the absence of sensor data during LAM mode, the burn is terminated by the on-board software logic and the spacecraft is normalized by ground intervention. But during mission critical operations, like in interplanetary missions, this logic fails to meet the mission requirements since there is no other opportunity to carry out the burn. In this context, a new fault tolerant intelligence approach is required. This chapter discusses the design and implementation of a new logic introduced in on-board software, to overcome LAM termination, in case of failure of sensor data updates. It also highlights the recovery time for various combinations of sensor failures.

M. Balaji · C. Mala (✉)
Department of CSE, National Institute of Technology, Tiruchirappalli, India
e-mail: mala@nitt.edu

M. Balaji
e-mail: bala.sure87@gmail.com

M. S. Siva
CDAD-I, CDSG, U R Rao Satellite Centre, ISRO, Bangalore, India
e-mail: mssiva@ursc.gov.in

© Springer Nature Switzerland AG 2020
J. Singh et al. (eds.), *A Journey Towards Bio-inspired Techniques in Software Engineering*,
Intelligent Systems Reference Library 185,
https://doi.org/10.1007/978-3-030-40928-9_2

2.1 Introduction

The Attitude Orbit Control System (AOCS) of a spacecraft provides a stable platform for its usage like imaging payloads. It provides the functionality of 3 axis attitude control with the help of thrusters in orbit raising. The Attitude Orbit Control Electronics (AOCE) receives the spacecraft attitude data from reliable sensors, spacecraft body rate from Dynamically Tuned Gyroscopes (DTG) and delivers output to the actuators. AOCE acts as an interface with sensors to measure attitude information, with actuators to maintain the desired attitude. This is a processor based system which interfaces with the other sub systems like Telecommand, Telemetry, and Payload etc. There is a redundant system always exists to carry if failure occurs in one system. All hardware elements including sensors are having redundancy in spacecraft. Due to the increase of requirement in certain projects like interplanetary missions, software complexity increases and software faults during operation also in-creases. The system is dependable system and relies on its availability, reliability, maintainability and integrity. To enhance the system robustness fault tolerance mechanism is required [1, 2]. Several techniques of fault tolerance schemes are available, based on the proper choice which suits for current configuration system it is to be chosen. N-Version programming is a technique where two hardware systems contains different version of software that does same operation. If one system fails to produce output other system is to be checked, if both the system fails to give proper output it is considered as failure. To overcome the catastrophic failure new design work in control algorithm is to be carried out for two systems and testing end to end is required for identifying faults if any. But this may not suit for mission critical applications. There are several methods like Checkpoint, Recovery Block, and Decision Making. Using backward recovery or checkpoint technique unpredictable errors can be handled but if error affects the recovery mechanism it's not considerable. In case of nested recovery without proper process synchronization it leads to domino effect [3]. Forward recovery is application specific and to be tailored to the system need and if the state is damaged beyond the threshold it cannot be used for recovery. Failure modes to be tested using injection of faults during testing by manipulating fault conditions and to verify the proper recovery is happening or not [4]. Software faults are mainly logical faults which failed to create a scenario of on-orbit in ground during development. Fault tolerance mechanism is highly effective if it is developed within the system with pro-active design features. Check point-ing mechanism is again a best fault tolerance scheme, when fault occurs it will retain from the last stored configuration [5] but it will be cumbersome to store entire configuration for the minor change in case of real time mission critical systems. Especially in case of crucial missions lot of on-board autonomy is required; LEB mode is taken in this chapter because launch vehicle places satellite in the intermediate orbit and from there burn is required to reach desired or-bit by LAM firing. This work is concentrated on the fault occurrence during burn or pre burn or in the post burn normalization. The duration of orbit transfers to be reduced [6] and also strategy should be carried out to limit the burn less than the least factor. On-board LEB computation using chebyshev approximations is essential in order to follow the steering reference along with body attitude to predict error that reduces ground calculation [7]. The effort is to reduce number of uplink coefficients from the ground and to calibrate steering output based on fitting algorithm using chebyshev polynomial approach is a prime task [8]. The total number of coefficients used in on-board to propose each item is 24 individual words of 32 bit [9, 10]. Likewise four individual Q components were used to derive reference Q's during burn, refer Annexure Table 2.3 for sample engineering values used for uplink. All geostationary, geosynchronous and interplanetary satellites use this polynomial based approach [11] to derive burn time attitude references. The accuracy of the burn coefficients were studied across the missions with ground simulation and with on-board data. Throughout the burn the spacecraft is corrected along with the profile reference and proper system path is maintained without external interactions. Post implementing of the proposed algorithm acceptance test is carried out. This test will acts as a low pass filter for the input data receives from the sensor select-ed, the aim is to discard all the data received from the sensor which is not in range or wrong data. Digital Sun Sensor (DSS) gives yaw error when sun crosses roll pitch plane. It tracks the sun in the range of $\pm64°$. DTG is a relative sensor gives two axes information with one redundancy unit. The current in the rotor spin measures the attitude change. Table 2.1 gives the sensor axis information for satellite. Star sensor gives information called sensor frame attitude in ECI frame. All these sensor information is validated using the proposed approach during burn. Fault test case to be injected in acceptance test to check the system availability after recovery.

2.2 Implementation

2.2.1 Fault Tolerance Intelligence Advisor System (FT-IAS)

LEB is a thrusters firing mode used in orbit raising of space-craft to achieve expected orbit, using sensors to derive attitude and actuators to perform firing. In case of orbit raising precise reference Q's for LEB is generated using on-board implementation logic. Up-link of LEB quaternion is a prerequisite for the on-board computation. Based on the duration provided in input as part of uplink, LEB Q reference is generated and the output is compared with spacecraft attitude at every 64 ms. Spacecraft attitude is derived from the selected sensor. Here DTG is a relative sensor used in loop to provide all three axes Yaw, Roll, Pitch information. Each sensor is redundant to provide change over if any fault occurs in the selected sensor. During LEB mode if any of the selected sensor fails in case of geo-stationary satellites LEB firing will be aborted, change over happen to put back the redundant sensor in to loop after sensor validation and ground intervention will be carried out to normalize the space-craft. Due to the failure of sensor in on-orbit, terminate of burn will be issued through on-board software and burn could not proceed further until next plan for burn is realized. In current scenario especially in inter-planetary mission(s), there is no

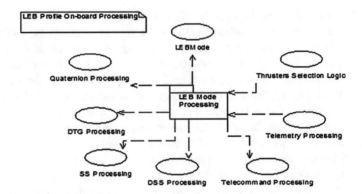

Fig. 2.1 UC-LEB profile mode

possible way to continue the burn when redundant sensor also fails or combination of all sensors fails there is no on-board logic exists to continue burn. It leads to single point failure in turn leads to mission failure. To overcome this, intelligence advisor system algorithm is proposed when fault occurs in any of the sensor failures during burn auto change over with all possible available sensors in an auto sequencer mode. Liquid Engine Burn Mode interacts with other modules called quaternion processing, DTG data processing, thruster's selection logic, telecommand processing and telemetry processing for its operation. DTG processing gives the quaternion of space-craft body to the controller, thrusters selection logic specifies the AOCE thrusters to be selected along with 440 N thrusters for orbit raising and stability correction of the spacecraft. The use case diagram depicts the inter-action between the other modules with LEB processing is shown in the Fig. 2.1. The implementation of this proposed logic in addition to the on-board software is considered by including computation load, periodicity, execution and memory usage of variables and constants used.

The block diagram representation of the proposed FT IAS approach is shown in Fig. 2.2. Nominally DTG is a prime relative sensor selected apriori for the control loop. Orbit reference will be given by LEB chebyshev model and spacecraft attitude (Qib) will be given by DTG sensors. Spacecraft error (Qerror) is derived based on the two outputs and given to controller to maintain stability. The sensor validation logic autonomously rejects the faulty sensor and change over to the next available sensor to continue the burn without abort or any interruption. The main activity of the LEB profile with proposed approach is shown in the Fig. 2.3. This computation is called in every major cycle (64 ms) to match with the controller algorithm execution frequency. LEB profile generator follows polynomial model and provides reference information to the processing module. The reliable sensors for attitude information

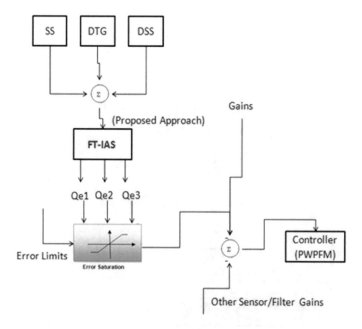

Fig. 2.2 Block diagram representation of the algorithm

using in FLIGHT model is DSS, DTG and SS. In this DTG is a relative sensor used to give attitude information during LEB and other modes. This FT-IAS is an adjudicator based approach embedded with on-board code to select the appropriate sensor during fault. During mission critical operations like orbit raising, fault can occur at anytime and it is unpredictable. Handling contingencies due to the sensors failure(s) during payload operations time is very much essential. Here all these sensors are connected with AOCE with common bus. Any fault occurred in sensors including bus failure or sensors processing failure, FT-IAS detects it and reconfigure to redundant sensor, assume redundant sensor is ON. Adjudicator mechanism controls and coordinates the sensor failures and its reconfiguration to redundant sensor without letting failures to stop running operation. The proposed algorithm contains three phases. LEB Profile Generator, QIB validation, Qerror derivation. Coefficients gets uplinked into on-board and profile generator computation looks for new coefficients and develop on-board LEB reference Q's by chebyshev polynomial approach. Qib is the spacecraft body Q's derived from sensors selected for the loop. If DTG is considered, this Qib will be updating by DTG sensor, if SS is selected Qib will be updating by star sensor. The adjudicator keep track of all these sensors selected and based on the updating frequency of the selected sensor processing it will detect any anomalies in the sensor data. If DTG1 is selected it will give two axes information called pitch and roll. Similarly DTG2 gives pitch and yaw, DTG3 gives roll and yaw axes information of

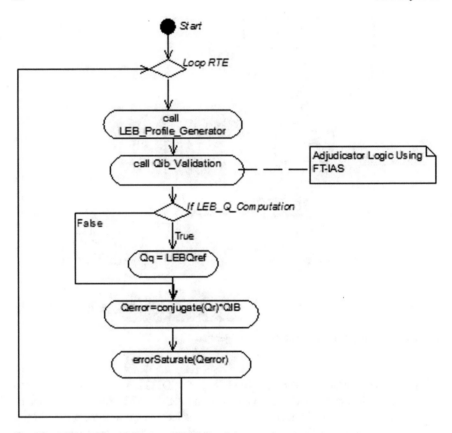

Fig. 2.3 LEB profile with proposed FT-IAS activity

spacecraft. At any given point of time any of the two DTG combinations gives all three axes data. DSS is a digital sun sensor gives two other axes information on its mounted plane. Star sensor gives all three axes information. In on-board logic all combinations derived from the proposed approach is designed and in case of any sensor failure auto change over takes place during burn. Intelligence advisor system (IAS) provokes alternate sensor autonomously in less than a second without affecting the burn progress. Qib validation logic is auto sequencer based and if logic triggers on-board it looks into correct sensor and provides uninterrupted burn progress that leads to proper attitude correction of spacecraft.

2.3   Algorithm

2.3.1   Proposed Algorithm: (Fault Tolerance Intelligence Advisor System During LEB Firing)

Algorithm 1 Proposed Algorithm: (Fault Tolerance Intelligence Advisor System during LEB Firing)

Preparation: Spacecraft and its sub subsystems with proper health condition are considered.
Initial Condition: Spacecraft mode is in LEB, LEB Q Com-putation is Enable, Pulse Width
Pulse Frequency Modulation (PWPFM) Enable, Rate Check Enable in all three axes
Elements Used: Qib, Qerror, Qr, LEBQref is initialize with zero.
Uplink LEB chebyshev polynomial coefficients to on-board memory (4 sets of 24 coefficients
each.)
Keep this Uplink LEB Q's as input for chebyshev model computation.
loop RTE do

BEGIN
call LEBProfileGenerator   //for LEB Termination and for LEB firing
call QibValidation   //Using FT-IAS
if (LEBQComputation) **then**
   $Qr \rightarrow$ LEBQref   //Computed by LEB Profile
end if
$Qerror \rightarrow Conjugate(\text{Qr}) * \text{Qib}$ //Qib update
Saturate Qerror   //Given to Controller Logic
   End loop RTE

2.4   Simulation and Results

The implementation of the algorithm is carried out in addition to the existing onboard software and tested in GSAT11 communication satellite setup. This software is written in ADA and embedded with AOCE, called as Real Time Executive (RTE) cycle. The onboard computer should complete all the scheduled tasks related to AOCE within 64 ms. Software is tightly coupled with the hardware and all the tasks are time sliced and scheduled. This scheduler algorithm is static, cyclic and generally non-interrupt based. The on-board soft-ware incorporates the mission requirement for chebyshev polynomial based LEB profile. This implementation is tested in real time em-bedded system setup using HX1750 processor working at a processor clock of 12 MHz. The processing time and calling frequency of each sensor processing and LEB profile generator is defined in the Fig. 2.4.

LEB attitude steering Q's output during burn with FT-IAS implementation is shown in Figs. 2.5, 2.6, 2.7 and 2.8 respectively. For each second approximately 15.6

Algorithm 2 QIB Validation

Sensor's used: DTG-3 Units. Each DTG gives 2 axes infor-mation.
a. DTG1 gives Pitch1, Roll1 Error information
b. DTG2 gives Pitch2, Yaw1 Error information
c. DTG3 gives Roll2, Yaw2 Error information (Redun-dant)
Digital Sun Sensor(DSS) gives redundant yaw axis infor-mation
Star Sensor(SS) gives all three axes information

CASE VALIDATION THEN
When 0
Initial case DTG sensor is considered for Qib Update
if (DTG_Status) **then**
$\quad Qib \leftarrow (Qib * Delta\,Q)$ //DeltaQ from DTG sensors
$\quad$ normalize Qib
else
$\quad validation \leftarrow 1$ //if Selected DTG fails
end if
When 1
Qib is invalid due to selected DTG failure. Change over to redundant DTG which is ON and SYNC
if (ChangeOver_DTG_Status) **then**
$\quad Qib \leftarrow (Qib * Delta\,Q)$ //DeltaQ from redundant DTG
$\quad$ normalize Qib
else
$\quad validation \leftarrow 2$ //if redundant DTG fails
end if
When 2
Consider left over DTG for control which gives two axes information and select appropriate DSS to give residue axis error information
if (LeftOver_DTG_Status)AND (DSS_Status) **then**
$\quad Qib \leftarrow (Qib * Delta\,Q)$ //DeltaQ from DTG + DSS
$\quad$ normalize Qib
else
$\quad validation \leftarrow 3$ //if left over DTG or DSS or both fails
end if
When 3
Using star sensor, update Qib. Assume SS is in healthy condition and SS is in TRACK mode.
Calculate $Qibas(Qib \leftarrow Qib * SSQib)$ //SS updates SSQib
normalize Qib
if (SS_Stats_Fails) OR (non_TRACK_mode) **then**
$\quad validation \leftarrow 0$ //Reset for next usage
$\quad$ abort LEB firing
$\quad$ Ground intervention required for burn replan
end if
When Others
NULL
END CASE

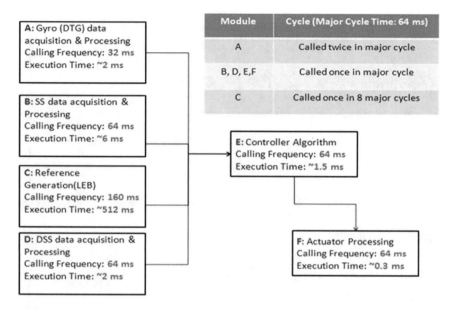

Module	Cycle (Major Cycle Time: 64 ms)
A	Called twice in major cycle
B, D, E,F	Called once in major cycle
C	Called once in 8 major cycles

A: Gyro (DTG) data acquisition & Processing
Calling Frequency: 32 ms
Execution Time: ~2 ms

B: SS data acquisition & Processing
Calling Frequency: 64 ms
Execution Time: ~6 ms

C: Reference Generation(LEB)
Calling Frequency: 160 ms
Execution Time: ~512 ms

D: DSS data acquisition & Processing
Calling Frequency: 64 ms
Execution Time: ~2 ms

E: Controller Algorithm
Calling Frequency: 64 ms
Execution Time: ~1.5 ms

F: Actuator Processing
Calling Frequency: 64 ms
Execution Time: ~0.3 ms

Fig. 2.4 Major cycle scheduler

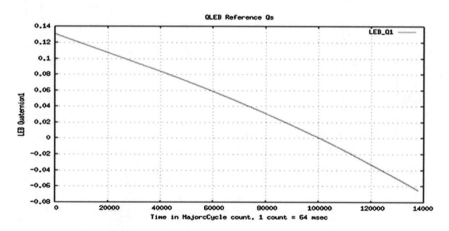

Fig. 2.5 LEB profile Q:Q1

times of software major cycle were executed. Normalized log is taken for every 64 ms of data in a faster sampling rate. It states that throughout the burn spacecraft steering attitude is derived and given to Qerror derivative for controller action. Qib up-dates from sensor will accumulate in calculation of Qerror along with LEB reference Q's. Qerror plot is shown in Figs. 2.9 and 2.10 and 2.11. At 8 s of Qerror computation DTG1 is declared as fault and change over happened to DTG3 is visible in the plot. At 15 s of instant DTG2 made failure and DSS along with DTG3 provides the attitude

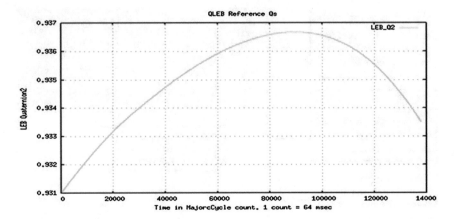

Fig. 2.6 LEB profile Q:Q2

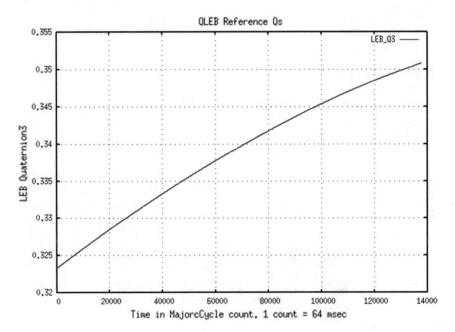

Fig. 2.7 LEB profile Q:Q3

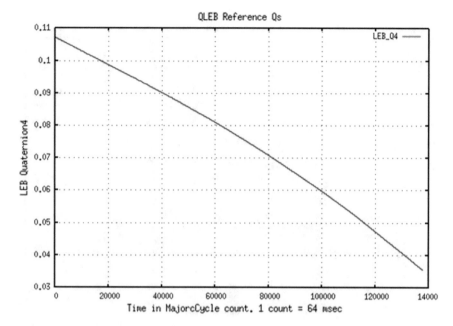

Fig. 2.8 LEB profile Q:Q4

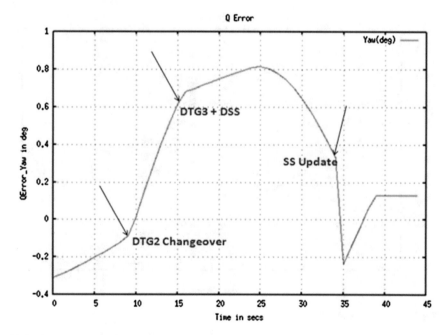

Fig. 2.9 Q error yaw rate

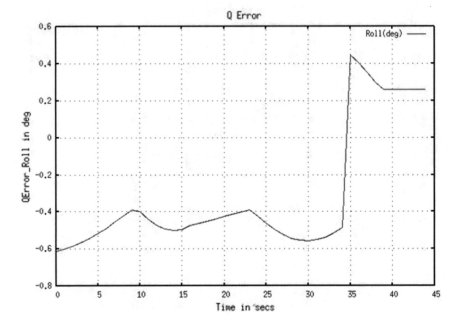

Fig. 2.10 Q error roll rate

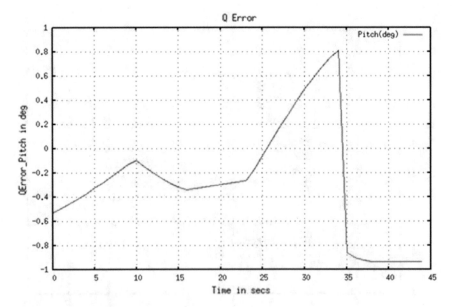

Fig. 2.11 Q error pitch rate

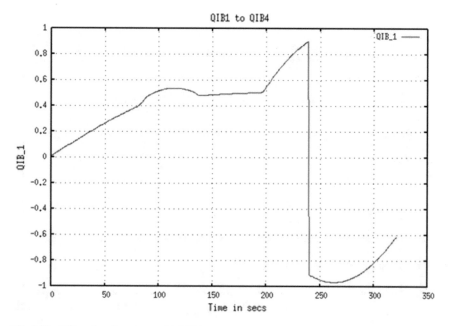

Fig. 2.12   QIB update from sensor(s)-QIB1

data refer throughout the burn Fig. 2.9. The roll error is specified in Fig. 2.10. The same is similar to yaw error and Fig. 2.11. Shows pitch error information. At 34 s of burn, DSS made fail and no more DTG sensor to give any axis information, to overcome the problem without terminate the secondary left out sensor called Star Sensor is selected into the loop. General assumption here is star sensor is to be ON and in TRACK mode, this to be ensured before start of the burn else in pre burn time make acquire star and keep star sensor in TRACK mode. Throughout the burn star sensor provides Qib information and there is no abort of LEB is traced. Qib's updated by different sensor is shown in the Figs. 2.12, 2.13, 2.14 and 2.15. Qib shows accumulated data throughout the update hence it is given in seconds; it covers of pre LEB mode, during LEB burn and post LEB mode. The proposed approach results in proper continuation of LEB burn with the help of maximum availability FLIGHT sensors used in on orbit to meet the project requirement by handling fault occurred during critical operations.

2.5   Conclusion and Future Work

Mission critical system is a high reliable dependant system; to avoid single point failure system hardware is always kept redundant and software in both the systems are identical which is designed conceptually and tested. Due to requirement changes

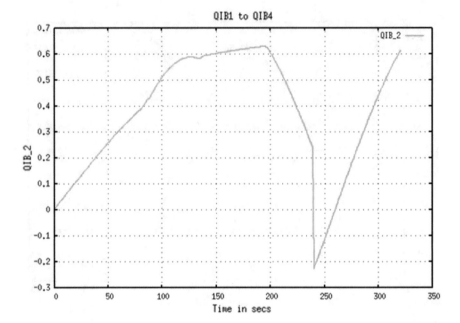

Fig. 2.13 QIB update from sensor(s)-QIB2

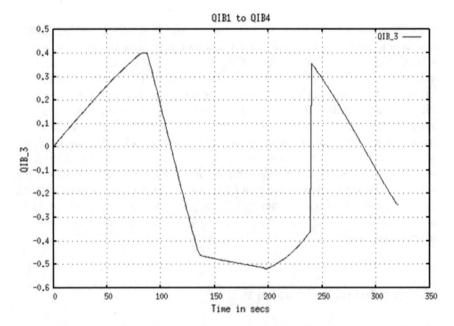

Fig. 2.14 QIB update from sensor(s)-QIB3

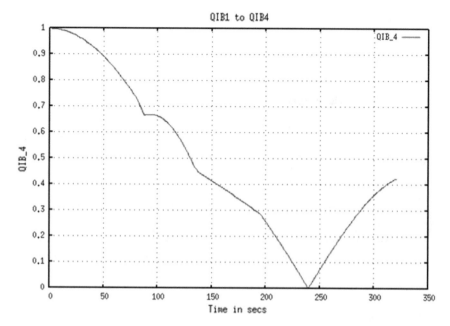

Fig. 2.15 QIB update from sensor(s)-QIB4

or increases in project to project, complexity of software also increases which in turn software faults during operation also increases. Here while considering interplanetary satellites orbit raising strategy is a crucial task, all possible sensors are to be used in loop to handle any unpredicted failure(s). At present FLIGHT on-board software lags of on-board autonomy to select different sensors in a sequencer mode within stipulated time without affecting the burn. It's a fault tolerance approach in this chapter we discussed to avoid LEB abortion or any critical operations fault and to continue burn with minimal delay of change over to the other redundant sensor's connected. The simulation results explain the change over time between the failures in an efficient way. By these results spacecraft attitude behavior is studied and it is in normal acceptable range. The proposed logic provides the efficient way of handling sensors autonomously with less mission interaction from ground like enabling or disabling the logic. In future this approach can be implement in ground simulation setup for interplanetary missions and intact with flown satellite condition, in prior to every burn test shall be carried out in simulation setup and to identify faults if any especially in orbit raising. It will reduce drastic amount of time spending in analyzing burn strategies prior to satellite orbit maneuvers.

Appendix

See Tables 2.1, 2.2, 2.3, 2.4.

Table 2.1 Sensor axis information

Sensor	Error axis Y-Yaw, R-Roll, P-Pitch	Processing execution time (in ms)
DTG1	P,R	
DTG2	P,Y	~2
DTG3	R,Y	
DSS1	Y	
DSS2	Y	~2
SS1	Y,R,P	
SS2	Y,R,P	~6

Table 2.2 Abbreviation

Name	Abbreviation
AOCE	Attitude Orbit Control Electronics
AOCS	Attitude Orbit Control System
DSS	Digital Sun Sensor
DTG	Dynamically Tuned Gyroscope
EPO	Elliptical Parking Orbit
FT-IAS	Fault Tolerance Intelligence Advisor System
LAM	Liquid Apogee Motor
LEB	Liquid Engine Burn
S/C	Spacecraft
SS	Star Sensor

Table 2.3 Sample engineering values considered for LEB steering profile

LEB Q1	LEB Q2	LEB Q3	LEB Q4
3.94678414e-02	9.34358418e-01	3.38355392e-01	7.37756193e-02
−9.72318947e-02	1.51092536e-03	1.38238641e-02	−3.56341787e-02
−6.92040008e-03	−2.01512291e-03	−1.24308944e-03	−2.66828830e-03
−8.36871855e-04	−2.89422111e-04	−7.34890418e-05	−3.16074846e-04
4.07293614e-04	−8.54941754e-05	−1.48642348e-05	1.21110003e-04
1.49957295e-05	2.15683940e-05	7.90562535e-06	−7.47336480e-06
7.38458184e-05	−3.34045103e-06	3.96096357e-06	3.76373755e-05
3.62617902e-05	−1.80680104e-06	3.25174096e-06	2.85307706e-05
−3.04157638e-05	3.87415002e-06	5.62686978e-07	−2.74519425e-06
−9.70267365e-07	−1.79950541e-06	−5.76037678e-07	−1.89090599e-07
−7.63145908e-06	5.05692412e-07	−1.03864772e-06	−1.02545437e-05
−1.16792971e-05	1.64789060e-06	−1.11446718e-06	−1.30171547e-05
7.27599217e-06	−6.53946017e-07	−4.57186246e-07	−3.32693116e-06
1.90139849e-06	1.16075924e-07	1.57118819e-07	1.29299906e-06
−4.00764549e-07	−4.92661245e-09	4.48026384e-07	4.60826732e-06
4.57010992e-06	−1.03697448e-06	6.32963861e-07	8.16083593e-06
−2.05675860e-06	3.64379957e-08	3.04023416e-07	2.73636147e-06
−1.67459734e-06	8.47248884e-08	−4.09118925e-08	−3.72037903e-07
1.27196517e-06	9.52592032e-08	−3.38272599e-07	−3.87766886e-06
−1.07694086e-06	5.74795138e-07	−4.08809655e-07	−5.35296886e-06
4.86733825e-07	3.46519698e-07	−3.47502834e-07	−4.21267077e-06
1.75566151e-06	−1.70888654e-07	3.06587133e-09	4.13232328e-07
−1.05672621e-06	6.19723366e-08	1.02241835e-07	1.19456126e-06
3.52195173e-07	−4.37235883e-07	3.16652176e-07	4.26824408e-06

Table 2.4 Sample reference output of LEB profile during burn

Time (in msec)	LEB_Q1	LEB_Q2	LEB_Q3	LEB_Q4
0.0000000000	0.1310204476	0.9310189248	0.3233402879	0.1071842441
0.0640000000	0.1310192591	0.9310190423	0.3233405589	0.1071838585
0.1280000000	0.1310180706	0.9310191599	0.3233408299	0.1071834729
0.1920000000	0.1310168820	0.9310192774	0.3233411008	0.1071830871
0.2560000000	0.1310156934	0.9310193950	0.3233413718	0.1071827013
0.3200000000	0.1310145047	0.9310195126	0.3233416428	0.1071823154
0.3840000000	0.1310133160	0.9310196302	0.3233419137	0.1071819293
0.4480000000	0.1310121273	0.9310197479	0.3233421847	0.1071815432
0.5120000000	0.1310109386	0.9310198655	0.3233424556	0.1071811569

(continued)

Table 2.4 (continued)

Time (in msec)	LEB_Q1	LEB_Q2	LEB_Q3	LEB_Q4
0.5760000000	0.1310097499	0.9310199832	0.3233427266	0.1071807706
0.6400000000	0.1310085611	0.9310201008	0.3233429975	0.1071803842
0.7040000000	0.1310073723	0.9310202185	0.3233432684	0.1071799976
0.7680000000	0.1310061834	0.9310203362	0.3233435394	0.1071796110
0.8320000000	0.1310049945	0.9310204539	0.3233438103	0.1071792243
0.8960000000	0.1310038056	0.9310205717	0.3233440812	0.1071788374
0.9600000000	0.1310026167	0.9310206894	0.3233443521	0.1071784505
1.0240000000	0.1310014277	0.9310208072	0.3233446230	0.1071780635
1.0880000000	0.1310002387	0.9310209250	0.3233448939	0.1071776764
1.1520000000	0.1309990497	0.9310210428	0.3233451648	0.1071772892
1.2160000000	0.1309978607	0.9310211606	0.3233454356	0.1071769018
1.2800000000	0.1309966716	0.9310212784	0.3233457065	0.1071765144

References

1. Ammann, P. E, Knight J C, Data Diversity: An Approach to Software Fault Tolerance. Proceedings of FTCS-17, IEEE Transactions on Computers, **37** 321–354 (1988)
2. Takano, T., Yamada, T., Kanekawa, N.: Fault-tolerance experiments of the Hiten onboard space computer, Digest of papers. In: Fault-Tolerant Computing: The Twenty-First International Symposium (1991)
3. Knight, J.C., Anderson, T.: A framework for software fault tolerance in real-time systems, vol. 9, pp. 55–364. Computing Laboratory, University of New-Castle-upon-Tyne (1983)
4. Dilenno, T.R., Yaskin, D.A., Barton, J.H.: Fault tolerance testing in the advanced automation system. In: IEEE Technical Symposium, USA (1991)
5. Huang, Y., Kintala, C.: Software Implemented Fault Tolerance: Technologies and Experience. AT&T Bell Laboratories, Murray Hill, NJ 07974
6. Koppel, C.R.: Low Thrust Orbit Transfer Optimiser for a Spacecraft Simulator 6th ICATT, Darmstadt (2016)
7. Deprit, A., et al.: Computation of ephemerides by discrete Chebyshev approximations. Navigation **26**(1), 1–11 (1979)
8. Sheela, B.V., Padmanabhan, P.: Compressed polynomial approach for onboard ephemeris representation. J. Guid. Control Dyn. **13**(4), 765–767 (1990)
9. Dakshayani, B.P., Rao, S.V.: Onboard ephemeris representation by a polynomial approach. J. Spacecr. Technol. **3**(2), 7–12 (2003)
10. Dakshayani, B.P., Nagamani, T., Satheesha, A.L., Ashwin, G.S., Ravikumar, L., Kiran, B.S.: Onboard Orbit and Attitude Representation Algorithm of IRNSS (2010)
11. Kiran, B.S., Gopinath, N.S., Negi, K.: MARS Orbiter mission design – Preliminary Analysis (2003)

Chapter 3
Type-2 Fuzzy Controller (T2FC) Based Motion Planning of Differential-Drive Pioneer P3-DX Wheeled Robot in V-REP Software Platform

Anish Pandey, Nilotpala Bej, Ramanuj Kumar, Amlana Panda and Dayal R. Parhi

Abstract In the present era, the wheeled robot performs various tasks like patrolling, disaster relief, and planetary exploration. For these tasks, a robust navigation algorithm is needed, which can autonomously drive the wheeled robot in any situations. Therefore, in this article, the authors try to design a Type-2 Fuzzy Controller (T2FC), which controls the steering angle based motion, direction, and orientation of the Differential-Drive Pioneer P3-DX Wheeled Robot (DDPWR) by using sensory information without any human intervention in different obstacle conditions. The multiple inputs (obstacles distances received from attached sensors) and single output (steering angle) T2FC have been taken for this purpose. Virtual Robot Experimentation Platform (V-REP) software based 3-dimensional simulation environment, and Graphical User Interface (GUI) based 2-dimensional simulation environment has used to show the motion control results of DDPWR by applying T2FC technique. The remote API functions of V-REP software have been used to make a connection between the MATLAB GUI and V-REP simulation. The developed MATLAB program handles the behavior of DDPWR in the V-REP software engineering platform. In addition, the comparison study has done between proposed T2FC technique

A. Pandey (✉) · N. Bej · R. Kumar · A. Panda
School of Mechanical Engineering, KIIT Deemed to be University,
Patia, Bhubaneswar 751024, India
e-mail: anish06353@gmail.com

N. Bej
e-mail: nilotpala2002@gmail.com

R. Kumar
e-mail: ramanujkumar22@gmail.com

A. Panda
e-mail: amlana.pandafme@kiit.ac.in

D. R. Parhi
Department of Mechanical Engineering, NIT Rourkela, Sundergarh 769008, India
e-mail: dayalparhi@gmail.com

© Springer Nature Switzerland AG 2020
J. Singh et al. (eds.), *A Journey Towards Bio-inspired Techniques in Software Engineering*,
Intelligent Systems Reference Library 185,
https://doi.org/10.1007/978-3-030-40928-9_3

47

with the previously developed Type-1 Fuzzy Controller to show the authenticity and robustness of the developed T2FC.

Keywords Type-2 Fuzzy Controller · Steering angle · Differential-Drive Pioneer P3-DX Wheeled Robot · Sensor · Virtual Robot Experimentation Platform

3.1 Introduction

Robots are helping the human from many decades to solve various problems and perform multiple tasks like transport, material handling, spray painting, cleaning, etc. During the navigation of any robot in any environment, it is very important to avoid collision with hurdles and move towards the target. Lin et al. [1] have designed and implemented the Type-2 Neural Fuzzy Controller tuned by Artificial Bee Colony (ABC) algorithm for controlling the motion of load-carrying mobile robots between the wall. Lin et al. [2] have implemented a Type-2 Neural Fuzzy controller hybridized with an evolutionary algorithm to train the Pioneer P3-DX wheeled robot for wall-following behavior. Castillo [3] has made comparison studies between Ant Colony Optimization (ACO) algorithm tuned Type-2 Fuzzy Logic, and Gravitational Search Algorithm (GSA) tuned Type-2 Fuzzy Logic and applied these for mobile robot torque control. The author stated that the ACO provided a better result as a torque controller. Pandey and Parhi [4] have proposed the hybrid Fuzzy Controller (FC) by combining Wind Driven Optimization (WDO) algorithm and used this hybrid FC for wheeled robot navigation and collision avoidance in different static and dynamic environments. In the article [5], authors have designed ultrasonic sensor and actuator controlled Adaptive Network Fuzzy Inference System (ANFIS) architecture to guide the differential-drive wheeled robot in the unknown environment between cluttered obstacles. Firefly Algorithm based motion planning of wheeled robot between uncertain environment has been presented in the article [6]. Chen and Liu [7] have shown the grid map based path planning of wheeled robot between partially known environment by applying Artificial Potential Field (APF) and ACO algorithm. In the article [8], the authors have worked on the position stabilization of wheeled mobile robot by applying Interval Type-2 Fuzzy Neural Network (IT2FNN). Ibraheem and Ibraheem [9] have proposed Modified Particle Swarm Optimization (MPSO) algorithm to tune and adjust the parameters of the nonlinear PID controller for trajectory tracking control of a differential drive two-wheeled non-holonomic robot. Singh and Thongam [10] have designed the sensor-actuator controlled multi-layer perceptron neural network for static and dynamic obstacle avoidance of a wheeled robot. In [11], the authors have implemented the Voronoi Diagram (VD) and Computation Geometry Technique (CGT) to control the motion and direction of a wheeled robot between moving obstacles.

After a critical literature survey, the author has found that the fuzzy logic techniques have been widely applied to motion control of a wheeled robot. However, most of the researchers have used Type-1 fuzzy for this purpose. Therefore, in this article, the author has tried to design a zero-order Takagi Sugeno based T2FC, which will control the steering angle of DGWR between the obstacles in different 2-D and 3-D scenarios. Rest of the article has organized in the following manners: Sect. 3.2 presents the brief about Type-2 Fuzzy Controller (T2FC) for Steering Angle Control of DGWR. V-REP and GUI Software based 2-D and 3-D Simulation Results and comparative study with the previously developed Type-1 Fuzzy Controller have been done in Sect. 3.3. Finally, Sect. 3.4 summarizes the Conclusion and Future Work.

3.2 Type-2 Fuzzy Controller (T2FC) for Steering Angle Control

This section presents a brief description of T2FC, which controls the steering angle based motion, direction, and orientation of the DDPWR. Figure 3.1 illustrates the developed model of Takagi–Sugeno Type-2 Fuzzy Controller for motion planning of DDPWR. The T2FC has five major components: fuzzifier, rule base system, fuzzy inference engine, type-reducer, and defuzzifier. The basic structure of the T2FC is shown in Fig. 3.2. The proposed T2FC has five major components: Fuzzifier, Inference engine, Rule-based system, Type-reducer, and Defuzzifier. The Fuzzifier coverts the given input to fuzzy value. In the rule base system, we specify 'if-then' rules by human knowledge. The inference engine makes the mapping between the given input to the output using fuzzy logic. The value of type-reducer is calculated by using mathematical integral equation listed in Eq. 3.4 below. The defuzzifier gives the single point value from the membership function by using weighted average method, center of area method, etc. The T2FC has multiple inputs that are obstacles distances received from attached sensors and has a single output that is steering angle of DDPWR. The name of the antecedent parameters of T2FC are Front Forward Obstacle Distance (FFOD), Left Forward Obstacle Distance (LFOD), and Right Forward Obstacle Distance (RFOD), respectively. Similarly, the consequent parameter is a Steering Angle (S_a). The antecedent part of T2FC selects two Gaussian membership functions, namely Near (Upper and Lower) and Far (Upper and Lower) that vary from 20 to 150 cm (taken from sensor data range). We use constant-type (zero-order Takagi Sugeno) fuzzy type-2 singleton membership function for consequent part of T2FC, which vary from −90 (negative steering angle) to 90 (positive steering angle). The three fuzzy type-2 singleton membership functions, namely Negative, Zero, and Positive, have been chosen for output (S_a) of T2FC. Figures 3.3 and 3.4 depict the Gaussian membership functions of inputs and constant-type (Takagi Sugeno) fuzzy type-2 singleton membership function for output, respectively. The user-defined if-then Type-2 fuzzy rule sets are given in Table 3.1, which control the movements of DDPWR during navigation in 2D and 3D simulation and experiments. The zero-order Takagi Sugeno type-2 fuzzy model [12] designs the proposed T2FC through the following equations:

$$Rule^i : \text{IF } u_1 \text{ is } M_1^i, \ u_2 \text{ is } M_2^i, \ \text{AND } u_3 \text{ is } M_3^i \text{ THEN } v \text{ is } f^i \qquad (3.1)$$

where $i = 1, 2, 3 \ldots 8$ (eight rules); in the above equation, u_1, u_2, and u_3 are represented the input variables FFOD, LFOD, and RFOD, respectively; and v indicates the output S_a. The M_1^i, M_2^i, and M_3^i are the interval of T2FC sets of the antecedent parameters, and f^i is the crisp value (number). The T2FC set M_j^i uses Gaussian membership function in the following forms:

$$\mu_j^i \left(u_j; c, \sigma \right) = \exp \left\{ -\frac{1}{2} \left(\frac{u_j - c_j^i}{\sigma_j^i} \right)^2 \right\} \qquad (3.2)$$

where $j = 1 \ldots 3$ (three input variables), the c_j^i and σ_j^i are tuning/adjusting parameters of the Gaussian membership function, known as centre and width, respectively. The width (σ_j^i) has been fixed and c_j^i is variable that is $c_j^i \in \left[c_{j1}^i, c_{j2}^i \right]$, where c_{j1}^i and c_{j2}^i are the centers of upper and lower membership functions of $\mu_j^i \left(u_j \right)$.

By applying the product inference function, the firing strength (δ^i) has calculated by the following function:

$$\delta^i = [\underline{\delta}^i, \bar{\delta}^i] = \left\{ \prod_{j=1}^n \underline{\mu}_j^i \left(x_j \right), \prod_{j=1}^n \bar{\mu}_j^i \left(x_j \right) \right\} \qquad (3.3)$$

where $\underline{\delta}^i$ and $\bar{\delta}^i$ are the firing strength of lower and upper bounds, respectively. The $\mu_j^i \left(u_j \right)$ uses $\underline{\mu}_j^i \left(u_j \right)$ and $\bar{\mu}_j^i \left(u_j \right)$ as lower and upper Gaussian membership functions, respectively of T2FC. After firing strength calculation, the type-reduced set is composed through the following equation:

$$V(x) = [v_l, v_r] = \int_{a_1} \ldots \int_{a_8} \int_{\delta^1 \in [\underline{\delta}^1, \bar{\delta}^1]} \ldots \int_{\delta^8 \in [\underline{\delta}^8, \bar{\delta}^8]} \frac{1}{\left(\frac{\sum_{i=1}^8 \delta^i \cdot f^i}{\sum_{i=1}^8 \delta^i} \right)} \qquad (3.4)$$

where the terms l and r denote the maximum and minimum bound values of v, respectively. If the values of δ^i and f^i are considered for v_l then these are specified by δ_l^i and f_l^i, respectively. Similarly, if the values of δ^i and f^i are considered for v_r then these are specified by δ_r^i and f_r^i, respectively. The defuzzification values of maximum (v_l) and minimum (v_r) bounds have been calculated by the weighted average method in the following forms:

$$v_l = \frac{\sum_{i=1}^8 \delta_l^i \cdot f_l^i}{\sum_{i=1}^8 \delta_l^i} \qquad (3.5)$$

$$v_r = \frac{\sum_{i=1}^8 \delta_r^i \cdot f_r^i}{\sum_{i=1}^8 \delta_r^i} \qquad (3.6)$$

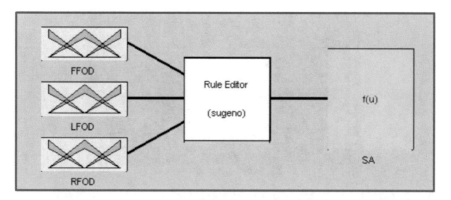

Fig. 3.1 The developed model of Takagi–Sugeno T2FC for motion planning of DDPWR

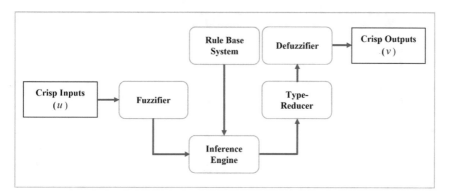

Fig. 3.2 Basic structure of the T2FC

The final defuzzification (or average of v_l and v_l) of the output v of T2FC as follows:

$$v = \frac{v_l + v_r}{2} \tag{3.7}$$

3.3 Computer Simulation Results

In this portion, the V-REP and GUI software-based simulation results have been presented to show the performance of T2FC for motion planning of DDPWR. Figure 3.5 illustrates that the Pioneer P3-DX robot is a differential drive two-wheeled robot and having sixteen ultrasonic sensors. These ultrasonic sensors can read obstacles from 2 to 400 cm. Two independent DC motor control the motion, direction, and orientation

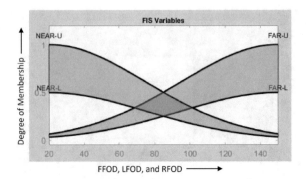

Fig. 3.3 Type-2 fuzzy membership functions for the inputs (FFOD, LFOD, and RFOD)

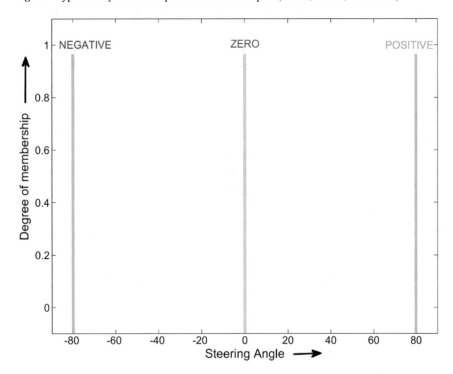

Fig. 3.4 Constant-type (Sugeno) fuzzy type-2 singleton membership function for output steering angle (S_a)

of DDPWR. One castor wheel, which is attached back side of DDPWR supports the chassis. The width of the DDPWR is 39 cm, the length is 44 cm, height is 23 cm, and its top speed is 1.2 m/s. The width and diameter of the wheel of DDPWR are 4.7 cm and 19.5 cm, respectively. The T2FC has three inputs (FFOD, LFOD, and RFOD) and single output steering control command to the DDPWR.

Table 3.1 Type-2 fuzzy rules set

Number	Rules
1	If (FFOD is Far) and (LFOD is Far) and (RFOD is Far) then (S_a is Zero)
2	If (FFOD is Near) and (LFOD is Near) and (RFOD is Near) then (S_a is Positive/Negative)
3	If (FFOD is Far) and (LFOD is Near) and (RFOD is Far) then (S_a is Positive)
4	If (FFOD is Far) and (LFOD is Far) and (RFOD is Near) then (S_a is Negative)
5	If (FFOD is Near) and (LFOD is Far) and (RFOD is Far) then (S_a is Positive/Negative)
6	If (FFOD is Near) and (LFOD is Near) and (RFOD is Far) then (S_a is Positive)
7	If (FFOD is Near) and (LFOD is Far) and (RFOD is Near) then (S_a is Negative)
8	If (FFOD is Far) and (LFOD is Near) and (RFOD is Near) then (S_a is Zero)

Fig. 3.5 Differential drive Pioneer P3-DX wheeled robot

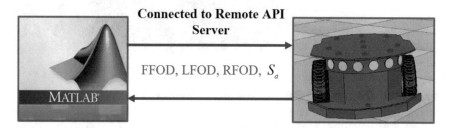

Fig. 3.6 The connection architecture between MATLAB and V-REP software

The remote API functions of V-REP software have used to make a connection be-tween the MATLAB GUI and V-REP simulation. Figure 3.6 shows the connection architecture between MATLAB and V-REP software, which sends the sensor-actuator control command to the DDPWR through remote API functions. We have

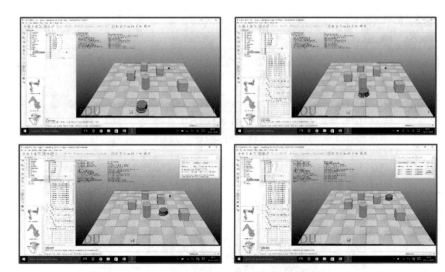

Fig. 3.7 3D simulation result of DDPWR between many obstacles using T2FC

Fig. 3.8 Recorded left and
right angular velocities of
motors of DDPWR in
Fig. 3.7

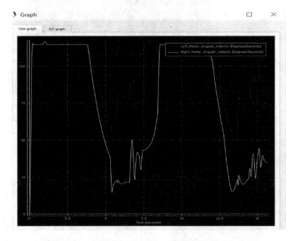

written the T2FC based DDPWR controlling program in MATLAB software language, and this program controls the motion of DDPWR in V-REP platform through
remote API functions. In the MATLAB program, we have declared if-else conditions like when the attached sensors find obstacles within the specified threshold
range, and then the designed T2FC provides the steering control command to the
DDPWR otherwise DDPWR navigates for goal. Firstly, the MATLAB receives the
sensor signals from V-REP, and these signals are feed to the T2FC as inputs then
T2FC gives desired steering angle control command as output for the DDPWR. This
desired steering angle has been sent to the DDPWR in V-REP platform through API
functions to control the motion of DDPWR for obstacle avoidance. The location of

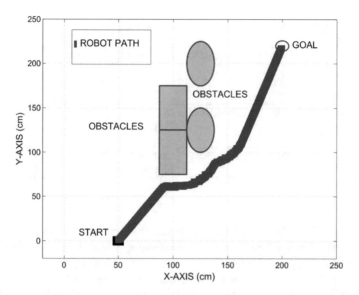

Fig. 3.9 2D simulation result of DDPWR between four obstacles using Type-1 Fuzzy Controller [13]

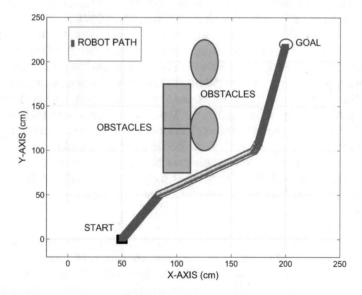

Fig. 3.10 2D simulation result of DDPWR between four obstacles using T2FC

Table 3.2 Comparison data between proposed T2FC and Type-1 Fuzzy Controller [13] in the same environment

Navigation controllers	Figure number	Start position	Goal position	Navigation path length (cm)	Final navigation path length error (cm)
Type-1 Fuzzy Controller [13]	Figure 3.9	(50 cm, 0 cm)	(200 cm, 220 cm)	115	3.77
T2FC	Figure 3.10	(50 cm, 0 cm)	(200 cm, 220 cm)	102	2.53

obstacles is unknown for the DDPWR, whereas the position of the goal is known for DDPWR. The V-REP platform provides the x, y, and z-axes locations of the DDPWR. Also, it gives the real-time navigation path length, time, and velocities. The 3D simulation result of DDPWR between many obstacles in V-REP software platform is given in Fig. 3.7. During the 3D simulation of DDPWR in Fig. 3.7, we have recorded the left and right motor angular velocities with respect to time in second, which is illustrated in Fig. 3.8. The red and yellow color lines indicate the left and right motor angular velocities (degree/second) of DDPWR.

Figures 3.9 and 3.10 illustrate the MATLAB based 2D simulation results of DDPWR between four obstacles using Type-1 Fuzzy Controller [13] and T2FC, respectively. Both the controllers have executed in the same environment, start, and goal positions. The start positions (50 cm, 0 cm) coordinates, goal positions (200 cm, 220 cm) coordinates, obstacle positions, and simulation environment have been taken for comparative analysis. In Fig. 3.9, the Type-1 Fuzzy Controller [13] controlled DDPWR has taken three turns from obstacles to reach the goal from starting position; whereas, in Fig. 3.10, the T2FC controlled DDPWR has taken two turns from obstacles to reach the goal from starting position. Table 3.2 shows the comparison data between proposed T2FC technique with the previously developed Type-1 Fuzzy Controller [13], which have been used for motion planning and obstacle avoidance. As we know, compared to type-1 fuzzy, the T2FC provides an additional degree of freedom to handle the uncertainties [8]. In the simulation results, we can see that T2FC has given smaller and smoother steering angle during obstacle avoidance due to its robustness against uncertainties. The comparison has been done by considering navigation path length (centimeter) and final navigation path length error (cm) from start position to the goal position. After summarizing the Table 3.2, we can say that the T2FC performed better and given a smooth trajectory compared to the Type-1 Fuzzy Controller [13]. Besides, we know that the less travel path length minimizes the consumption of energy; and the proposed T2FC has covered the short distance to reach the goal that means our controller is also better in terms of energy-saving.

3.4 Conclusion and Future Work

In this article, the authors have proposed T2FC to control the motions of DDPWR autonomously between obstacles. The MATLAB software-based program has handled the DDPWR in V-REP software engineering platform through the remote API function. As compared to previously developed Type-2 Fuzzy [13], we have used less number of rules (only eight rules) in this study and found good navigation results in different 2D and 3D platforms. Also, the comparative study has been done with the previously developed Type-1 Fuzzy Controller [13] to show the robustness of the developed T2FC. Both the 2D and 3D simulation results using MATLAB and V-REP software on Pioneer P3-DX robot are presented to show the effectiveness of the T2FC. In future, the Gaussian membership function of T2FC will be optimized and tuned by evolutionary algorithms.

References

1. Lin, C.H., Wang, S.H., Lin, C.J.: Interval type-2 neural fuzzy controller-based navigation of cooperative load-carrying mobile robots in unknown environments. Sensors 18, 1–22 (2018)
2. Lin, T.C., Chen, C.C., Lin, C.J.: Navigation control of mobile robot using interval type-2 neural fuzzy controller optimized by dynamic group differential evolution. Adv. Mech. Eng. 10, 1–20 (2018)
3. Castillo, O.: Bio-inspired optimization of type-2 fuzzy controllers in autonomous mobile robot navigation. Advanced Control Techniques in Complex Engineering Systems: Theory and Applications, vol. 203, pp. 187–200 (2019)
4. Pandey, A., Parhi, D.R.: Optimum path planning of mobile robot in unknown static and dynamic environments using fuzzy-wind driven optimization algorithm. Def. Technol. 13, 47–58 (2017)
5. Subbash, P., Chong, K.T.: Adaptive network fuzzy inference system based navigation controller for mobile robot. Front. Inf. Technol. Electron. Eng. 20, 141–151 (2019)
6. Patle, B.K., Pandey, A., Jagadeesh, A., Parhi, D.R.: Path planning in uncertain environment by using firefly algorithm. Def. Technol. 14, 691–701 (2018)
7. Chen, G., Liu, J.: Mobile robot path planning using Ant Colony Algorithm and Improved Potential Field method. Comput. Intell. Neurosci. Article ID 1932812, 1–10 (2019)
8. Kim, C.J., Chwa, D.: Obstacle avoidance method for wheeled mobile robots using interval type-2 fuzzy neural network. IEEE Trans. Fuzzy Syst. 23, 677–687 (2015)
9. Ibraheem, G.A., Ibraheem, I.K.: Motion control of an autonomous mobile robot using modified particle swarm optimization based fractional order PID controller. Eng. Technol. J. 34, 2406–2419 (2016)
10. Singh, N.H., Thongam, K.: Neural network-based approaches for mobile robot navigation in static and moving obstacles environments. Intell. Serv. Robot. 12, 55–67 (2019)
11. Ayawli, B.B.K., Mei, X., Shen, M., Appiah, A.Y., Kyeremeh, F.: Mobile robot path planning in dynamic environment using Voronoi diagram and computation geometry technique. IEEE Access 7, 86026–86040 (2019)
12. Juang, C.F., Hsu, C.H.: Reinforcement interval type-2 fuzzy controller design by online rule generation and Q-value-aided ant colony optimization. IEEE Trans. Syst. Man Cybern. Part B (Cybern.) 36, 1528–1542 (2009)
13. Cherroun, L., Boumehraz, M.: Fuzzy behavior based navigation approach for mobile robot in unknown environment. J. Electr. Eng. 13, 1–8 (2013)

Chapter 4
An Object-Oriented Software Complexity Metric for Cohesion

A. Joy Christy and A. Umamakeswari

Abstract The quality of object oriented (OO) software is often measured with the help of OO metrics—A specific type of software metrics that particularly evaluates the features of OO programming such as modularity, understandability, readability, reusability and extensibility. OO metrics that are derived by the concepts of cohesion and coupling depict good measures of module design in such way the module that shares high quotient of data within the module is better than sharing data between them. Thus, the quantification of data/information sharing within the software module is still in the thrust area of research for the approval of module design. The adaptability of existing cohesion metrics is a major concern for the developer as the metric value does not define a clear boundary between low, medium and high cohesions. To overcome this issue, in this chapter, a novel cohesion metric called Cohesion in Method (CohM) is proposed. The metric clearly defines a boundary between the levels of cohesionsby taking the benefits of Jaccard similarity measure. Thus, helps to evaluate software for reusability and maintainability. The results show that the metric comparatively gives better results than the traditional cohesion metrics by clearly differentiating medium and high level cohesions in software module.

4.1 Introduction

The term software metric is a quantifiable measure for validating the performance of software products in quality perspectives [1]. The promising features of object oriented programming from complexity to reusability, from maintenance to maintenance effort prediction inspires the development of OO software in IT industries than the traditional procedure oriented programming [2]. Modularity is one of the most vital characteristics of OO paradigm that fragments the large task into small well-

A. Joy Christy · A. Umamakeswari (✉)
School of Computing, SASTRA Deemed to be University, Thanjavur, India
e-mail: umamakeswari.arumugam@gmail.com

A. Joy Christy
e-mail: joychristy@cse.sastra.edu

© Springer Nature Switzerland AG 2020   59
J. Singh et al. (eds.), *A Journey Towards Bio-inspired Techniques in Software Engineering*,
Intelligent Systems Reference Library 185,
https://doi.org/10.1007/978-3-030-40928-9_4

defined units called modules. A module of an OO programming consists of methods that operate on same data relating to a single objective [3, 4]. Good design of a module clearly explains its contribution to the overall system there by eases the processes of readability and understandability for consequent modification and maintenance of software. The design of OO modules are evaluated by two vital parameters called cohesion and coupling where the prior measures the strength of relationship between the methods of module and the posterior measures the inter-relatedness across the methods of modules. Software metrics on cohesion and coupling always holds a stagnant position in evaluating the efficiency of software modules with a fact that the software with high cohesion and low coupling is good than its inverse [5].

Many cohesion metrics have been suggested in the history of object oriented programming. But, none of those metrics are quite popular as cyclomatic complexity metric as there is no work instantiated to prove the best cohesion metric ever since [6]. Moreover, the analysis on the existence of cohesion in OO software is highly challengeable, since the traditional metrics are lacking in describing its level of complexity such as low, medium and high rather they only indicate the presence of cohesion as a discrete value which is unable to interpret. This is indeed a problem for the software testers to decide upon the approval of the OO module. Thus, the research hypothesis questions can be set as:

- Is there a metric that clearly depicts the cohesion complexity of OO modules?
- Is there a cohesion metric that simplifies the decision on the approval/disapproval of OO modules?
- Is there a cohesion metric that can be proven as best and widely adopted by the software testers like cyclomatic complexity metric?

In this conjecture, this chapter proposes a new cohesion metric called CohM for identifying the level of cohesion in software module by resulting a value ranging between the values 0–1 to estimate the complexity of modules. CohM value 0 represents high cohesion, 1 for low and the intermediate values denotes medium cohesion. The metric employs Jaccard similarity measure for evaluating the cohesion level. The approval/disapproval of the module can be decided by executing an if-then decision as follows.

- High cohesion—approve the module (equals to 0—low complexity)
- Medium cohesion—make changes to obtain high cohesion and approve module (>0 and <1-medium complexity)
- Low cohesion—Disapprove module(equals to 1—High complexity)

Note: The disapproval of module necessitates the splitting of module into two or more to bring in cohesion. None of the cohesion metric results value from 0 to 1 indicating high, medium and low cohesions. Thus, the metric helps the tester to assess the level of cohesion and to accept/reject the module. Since, the metric clearly defines the boundaries on cohesion level, it can be widely adopted in software testing field. The cohesion metrics have been implemented as a software tool to conduct the theoretical, experimental validation of the proposed metric and to compare the performance with other cohesion metrics. The remaining sections of the chapter

are organized as follows: Sect. 4.2—literature review, Sect. 4.3 CohM—A Cohesion
Complexity Metric, Sect. 4.4—CohM Calibration, Sect. 4.5—Theoretical Validation
of CohM and Sect. 4.6—conclusion.

4.2   Review of Literature

Lack of Cohesion in Methods (LCOM1), the software metric proposed by Chidamber
and Kemerer [7, 8] is recognized as the base metric for any novel cohesion metric
in object oriented programming. The metric elucidates the number of method pairs
that do not share any instance variables within them. LCOM1 is simple but lacks in
the differentiation among good and bad cohesions. LCOM2 is an improvised version
of LCOM1 that computes the intersected and non-intersected methods of a module
with an implication that the data members shared by more number of methods are
relevant and increases module cohesion. Every method in a module is intersected
with every other methods of same module and increments P for null intersections or
Q for not-null intersections. The positive value of LCOM2 obtained by the difference
of P and Q denotes a high degree of null intersections which is a negative indication
of module design. On the other hand, zero and other negative values (set to 0) indicate
that there exists some amount of cohesion in software modules. The mathematical
notation of LCOM2 is defined as the following:

Consider a software module S with 'n' methods $m_1 \ldots m_n$ each of which denoted
by a set I_1. I_n respectively, consisting the data members used by them. Let P and Q
be the number of null and not-null intersections formed by the methods and LCOM2
is the difference between P and Q as shown in Eq. 4.1:

$$LCOM2 = || - |Q|, if \ |P| > |Q| \ or \ 0 \ otherwise \qquad (4.1)$$

LCOM2 is claimed as an effective measure for identifying the cohesion in software
modules, but not resilient with the elucidation of different levels of cohesions such
as low, medium and high. Moreover, the scope of LCOM2 is not extended to the
inherited modules. Li et al. [9] instead proposed LCOM3, a new graphical definition
for LCOM2 as it gives the value of zero for different level of cohesion classes.
LCOM3 is the fraction of methods that do not access a specific instance variable with
the overall variables in the class. The mathematical notation of LCOM3 is depicted
in Eq. 4.2.

$$LOM3 = 1 - \frac{sum(mV)}{m * v} \qquad (4.2)$$

where,
m refers to the number of class methods.
v refers to the number of variables in class.
mV refers to the methods that uses the variable.

sum (mV) refers to the sum of mV variables of a class.

LCOM2 fails to process and returns 0, if the number of methods or variables in the class is 0.

Bieman et al. [10] stated that the traditional LCOM indicates the lack of cohesion corresponds to direct communication between methods and proposed two OO metrics for measuring cohesion in class such as Tight Class Cohesion and Loose Class Cohesion. The metrics exhibit the direct and indirect sharing of instance variables with in pair of methods. TCC is observed as the number of directly connected methods and LCC is observed as the number of direct or indirect communications of components which are defined in the following Eqs. 4.3 and 4.4:

$$TCC(C) = \frac{NDC(C)}{NP(C)} \tag{4.3}$$

$$LCC(C) = \frac{NDC(C) + NIC(C)}{NP(C)} \tag{4.4}$$

where,

NDC refers to the count of direct communications.

NP refers to the maximum possible connections with class methods akan $x(n-1)/2$.

NIC refers to the count of indirect connections in an abstract class (AC(C)).

An abstract class is a collection of Abstract Methods AMs where each AM is a visible method in the class. Equation 4.5 denotes the mathematical notation of an AC:

$$AC(C) = AM(M)|M \epsilon V(C) \tag{4.5}$$

where,

M refers to the count of methods in a class.

V(C) refers to the number of visible methods in class (C) and the ancestor classes of C.

Here, AM refers to an abstracted method, defined with set of instance variables that can directly or indirectly be used in the methods of a class shown in Eq. 4.6

$$AM(M) = DU(M) \cup IU(M) \tag{4.6}$$

where,

DU(M) is a set with instances directly used by method M.

IU (M) is a set with instances indirectly used by method M.

High TCC and LCC value denotes high cohesion and vice versa. The metric proves that the implications of inheritance have negative impact with cohesion.

Vijayaraj et al. [11] have combined the advantages of cohesion and cognitive metrics to identify the cohesion type hidden in the module. A module with high cohesion is highly preferable than a module with low cohesion. Thus, the proposed

CCMCF assess the existence of cohesion through cognitive weights. The CCMCF can be calculated using Eq. 4.7.

$$CCCMF = \frac{\sum_n^{i=1} cognitiveweights(V_i)}{fv} \tag{4.7}$$

where, CCMCF is the Cognitive complexity metric for cohesion factor Cognitive of Weights (Vi) is the weight of variables 1 n, when the variables occurs for the first time in a method, the weight is one and gradually increases. If a variable is used multiple times in a method, only one copy of the variable is taken.

f refers to the count of module functions.

v refers to the count of module variables.

4.3 CohM—A Cohesion Complexity Metric

The denial on the verification of cohesion in software modules implicit the negligence of quality in software code, as high cohesion ensures reliability, understandability and modifiability of software. LCOM does not classify the levels of complexity in cohesion such as high, medium and low. This analysis may help the programmer to verify the quality of program code as an effort for software maintenance. Thus, this chapter proposed a novel approach to evaluate the complexity of modules with cohesion. Cohesion ensures the sharing of variables between the methods of a module with an assumption that the methods which share variables are relevant and meaningful. The more the variables are shared is more the cohesion is. If the methods are independent and do not share the variables in a module, then that can be split to form new classes. LCOM metric computes the degree of dependence by intersecting a method with every other method. In this chapter, a Jaccard distance measure is used to compute the degree of cohesion by quantifying the similarity of methods in terms of shared variables. The proposed metric can be denoted using the Eq. 4.8.

$$cohM = \frac{\sum_{i=1}^n \sum_{j=i+1}^n |1 - \frac{i \cap j}{i \cup j}|}{\frac{n(n-1)}{2}} \tag{4.8}$$

where, CohM refers to the proposed metric for cohesion.

n refers to the count of module methods.

CohM is the fraction sum of Jaccard distance of a method with all other methods by the number of methods in the module.

4.4 CohMmetric Calibration

The methodology of the proposed CohM metric is analyzed through an illustration. This study is conducted with three object-oriented programs inferring the three complexity levels of cohesion such as low, medium and high. The programs have been taken from the experiments that were performed in the OO programming laboratory with an objective to prove the high complexity programs gets high CohM value and low complexity programs gets lowCohM value. The code for the experimental programs is shown Table 4.1:

Table 4.1 Experimental programs

Program1.cpp: Low Complexity	Program2.cpp: Medium Complexity	Program3.cpp: High Complexity
class Rectlow	class Rectmedium	class Recthigh
{	{	{
int length, width;	int length, width;	int length, width,a,b,c;
float area;	float area;	float area;
void fm1 ()	void sm1()	void tm1()
{	{	{
cout << enter length<< endl;	cout<< enter length <<endl;	cout<< enter length <<endl;
cin>>length;	cin>>length;	cin>>length;
cout<< enter width<<endl;	cout<< enter width <<endl;	cout<< enter width <<endl;
cin>>width;	cin>>width;	cin>>width;
area=length * width	}	area=length*width;
}	void sm2()	cout<< the area is <<area;
void fm2()	{	}
{	area=length*width;	void tm2()
cout<< the length is << length;	}	{
cout<< the width is<< width;	void sm3()	a=10;
cout<<the area is<< area;	{	b=20;
}	cout<< the length is <<length;	c=a*b;
};	cout<< the width is <<width;	cout<< the multiplication of variables a and b is <<c;
void main()	cout<< the area is <<area;	}
{	}	};
Rectlow r;	};	void main ()
r.rect ();	void main()	{
r.putdata();	{	Recthigh r;
}	Rectmedium r;	r.rect();
	r.getdata();	r.multiply();
	r.calculate();	}
	r.putdata();	
	}	

Program1.cpp

Since, Program1.cpp has only two methods, the Jaccard similarity of the method is calculated for those methods as follows:
fm1()={length,width,area}
fm2()={length,width,area}

$$CohMm_{Program1.cpp} = \frac{1 - \frac{fm1() \cap fm2()}{fm1() \cup fm2()}}{\frac{n(n-1)}{2}} = \frac{1 - \frac{3}{3}}{\frac{2(2-1)}{2}} = \frac{0}{1} = 0 \qquad (4.9)$$

Program2.cpp
Sm1 () = {length,width}
Sm2 () = {area, length,width}
Sm3 () = {area,length,width}

$$iteration_{(1,2)} = 1 - \frac{sm1() \cap sm2()}{sm1() \cup sm2()} = 1 - \frac{2}{3} = \frac{1}{3} = 0.33 \qquad (4.10)$$

$$iteration_{(1,3)} = 1 - \frac{sm1() \cap sm3()}{sm1() \cup sm3()} = 1 - \frac{2}{3} = \frac{1}{3} = 0.33 \qquad (4.11)$$

$$iteration_{(2,3)} = 1 - \frac{sm2() \cap sm3()}{sm2() \cup sm3()} = 1 - \frac{3}{3} = \frac{3}{3} = 1 \qquad (4.12)$$

Sum of all iterations is 0.666.

$$CohMm_{Program2.cpp} = \frac{0.666}{\frac{n(n-1)}{2}} = \frac{0.666}{3} = 0.222 \qquad (4.13)$$

Program3.cpp
tm1 () = {area, length, width}
tm2 () = {a, b, c}

$$iteration_{(1,2)} = 1 - \frac{tm1() \cap tm2()}{tm1() \cup tm2()} = 1 - \frac{0}{6} = 1 - 0 = 1 \qquad (4.14)$$

$$CohMm_{Program3.cpp} = \frac{1}{\frac{n(n-1}{2}} = \frac{1}{1} = 1 \qquad (4.15)$$

The results of the three experimental programs shown in Table 4.2 reveals that the proposed CohM metric outputs 0 (High Cohesion) when all the variables are shared among the methods of the class, 1 (Low Cohesion) when the methods are not sharing the any of the variables and 0.222 (Medium Cohesion) when there is a partial sharing of variables between them. On the other hand, the outputs of LCOM1 and LCOM2 for the same programs are 0, 0 and 1 in which the interval between high and medium cohesions are not well differentiated. Though the metrics LCOM3, TCC and LCC

Table 4.2 Comparative analysis of cohesion metrics

Program name	CohM value	LCOM1	LCOM2	LCOM3	TCC	LCC
Program1.cpp	0	0	0	0	1	1
Program2.cpp	0.222	0	0	0.11	0.666	0.666
Program3.cpp	1	1	1	0.5	0	0

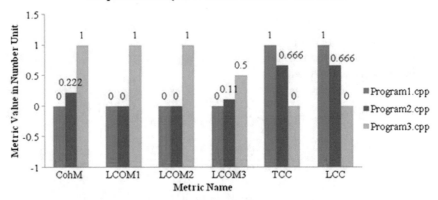

Fig. 4.1 Metric value analysis CohM versus Cohesion Metrics

overcomes the drawbacks of LCOM1 and LCOM2 it does not depict the level of cohesion complexity of the illustration programs.

Thus, it has been observed that the proposed CohM metric may be suitable for evaluating the cohesion in modules. The pictorial representation of the metric values is depicted in Fig. 4.1.

4.5 Theoretical Validation of CohM

The strength of any software metric can be evaluated through theoretical and empirical validations [12]. Weyukers properties are used as a standard theoretical validation method for assessing the strength of traditional metrics in object oriented programming [13–16], and [17]. Thus, in this chapter, Weyukers validation measure is used for the theoretical validation of proposed CohM metric. The empirical validations of the software metric can be demonstrated using experiments, case studies and surveys. Theoretical Validation: Weyukers Properties

Property 1
Non-coarseness: Given three modules A, B and C, their corresponding metric values μA,μB and μC, found such that

$$\mu A \neq \mu B \neq \mu C.$$

CohM does not measure all modules as equally as complex. CohM metric values for the experimental programs demonstrated in the CohM calibration are different from one another and does not result same value for all modules.

Property 2

Granularity: There can be n finite number of classes having same metric value. CohM may result the same metric value for n finite number of classes if the design implementation of those modules are same.

Property 3

Non-uniqueness: There can be two distinct modules A and B with a same metric value such that

$$\mu A = \mu B.$$

CohM results same metric values for classes A and B if they are equally complex.

Property 4

Design Details Important: Two modules A and B having same functionality do not imply

$$\mu A = \mu B.$$

If two modules A and B performing the same tasks are different, CohM value will also be different.

Property 5

Monotonicity: When two modules A and B are combined, the metric values of the two modules should be

$$\mu A \leq \mu(A + B) \, and \, \mu B \leq \mu(A + B)$$

CohM metric value for the combined classes will be greater than the individual modules A and B, if and only if the module share a high quotient of variable sharing among the methods of modules.

Property 6

Nonequivalence of Interaction: Interaction of Module C with two modules A and B having values $\mu A = \mu B$ does not imply

$$\mu(A + B) = \mu(B + C).$$

CohM metric value will be different from each other if the interaction between modules A and C, B and C are different and depends upon the degree of fitness with class C with A and Class C with B.

Property 7

Permutation: Permutation of modules code may impact metric value. Since, the permutation of functions do not change value, this property is not applicable for OO programming.

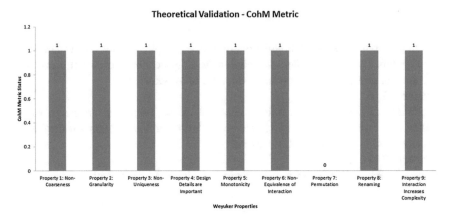

Fig. 4.2 Theoretical validation of CohM metric

Property 8
Renaming: Renaming of module A do not change value A. Renaming of modules does not affect cohM metric value.

Property 9
Interaction Increases Complexity:
    The complexity of a module (A + B) derived from two modules A and B increase interaction as

$$\mu(A) + \mu B \leq \mu(A + B).$$

CohM metric value of the combined module is higher than the sum of metric values of individual modules A and B as the interaction increases complexity in modules.

    The summary of CohM metrics theoretical validation against Weyukers Properties is presented in Fig. 4.2. Where, value 1 denotes that the properties are satisfied by CohM metric and 0 denotes that the properties are not suitable for OO metrics.

    The metric that satisfies at least seven properties of Weyukers is proven to be valid. Since, the CohM metric satisfies 8 properties, the metric is theoretically proven to be valid.

4.6 Conclusion

The role of software metrics is immense in bringing out quality in software programming. High cohesion in OO programming ensures in integrity of methods within modules, this would result in ease of maintenance of software. A software module with high cohesion is understandable, comprehendible and reusable. The existing LCOM metric does not differentiate high and medium cohesion separately. Thus, LCOM lacks in the insight speculation of cohesion in modules. To overcome this issue, in this chapter cohesion metric CohM is proposed. CohM verifies the cohesion complexity in software module, where the value 0 denotes high cohesion, 1 denotes low cohesion and the intermediate values denote partial cohesion. It has been suggested that the software module with high cohesion (value 0) can immediately be approved by the testers than the medium and low cohesive modules. A module with CohM value between 0.5 and 1 can be modified to bring in more cohesiveness within the module by incorporating relevant methods or separating irrelevant methods.

References

1. Boehm, B.W., Brown, J.R., Lipow, M.: Quantitative evaluation of software quality. In: Proceedings of the 2nd International Conference on Software Engineering, pp. 592–605. IEEE Computer Society Press (1976)
2. Asagba, P.O., Ogheneovo, E.E.: A comparative analysis of structured and object-oriented programming methods. J. Appl. Sci. Environ. Manag. 12(4) (2008)
3. Athanasopoulos, D., Zarras, A.V., Miskos, G., Issarny, V., Vassiliadis, P.: Cohesion-driven decomposition of service interfaces without access to source code. IEEE Trans. Serv. Comput. 8(4), 550–562 (2015)
4. Al Dallal, J., Morasca, S.: Predicting object-oriented class reuse-proneness using internal quality attributes. Empir. Softw. Eng. 19(4), 775–821 (2014)
5. Kalantari, S., Alizadeh, M., Motameni, H.: Evaluation of reliability of object-oriented systems based on Cohesion and Coupling Fuzzy computing. J. Adv. Comput. Res. 6(1), 85–99 (2015)
6. Etzkorn, L.H., Gholston, S.E., Fortune, J.L., Stein, C.E., Utley, D., Farrington, P.A., Cox, G.W.: A comparison of cohesion metrics for object-oriented systems. Inf. Softw. Technol. 46(10), 677–687 (2004)
7. Chidamber, S.R., Kemerer, C.F.: A metrics suite for object oriented design. IEEE Trans. Softw. Eng. 20(6), 476–493 (1994)
8. Hitz, M., Montazeri, B.: Chidamber and Kemerer's metrics suite: a measurement theory perspective. IEEE Trans. Softw. Eng. 22(4), 267–271 (1996)
9. Li, W., Henry, S.: Object-oriented metrics that predict maintainability. J. Syst. Softw. 23(2), 111–122 (1993)
10. Bieman, J.M., Kang, B.K.: Cohesion and reuse in an object-oriented system. ACM SIGSOFT Softw. Eng. Notes 20(SI), 259–262 (1995)
11. Vijayaraj, N., Ravi, T.N.: An advanced cognitive complexity metric for cohesion factor. Int. J. Pure Appl. Math. 118(9), 85–90 (2018)
12. Weyuker, E.J.: Evaluating software complexity measures. IEEE Trans. Softw. Eng. 14(9), 1357–1365 (1988)
13. Aggarwal, K.K., Singh, Y., Kaur, A., Malhotra, R.: Software design metrics for object-oriented software. J. Object Technol. 6(1), 121–138 (2007)
14. Anbumani, K., Srinivasan, K.P.: A set of object-oriented design metrics. J. Inst. Eng. 86, 1–9 (2005)
15. Chidamber, S.R., Kemerer, C.F.: A metrics suite for object-oriented design. IEEE Trans. Softw. Eng. 20(6), 476–493 (1994)
16. RadhikaRaju, P., AnandaRao, A.: A metrics suite for variable categorization to support program invariants. Int. J. Softw. Eng. Appl. 5(5), 65–83 (2014)
17. Rajnish, K.: Theoretical validation of inheritance metrics for object-oriented design against briands property. Int. J. Inf. Eng. Electron. Bus. 3, 28–33 (2014)

A. Joy Christy is working as Assistant Professor in School of Computing at SASTRA Deemed University, Thanjavur. She completed her Ph.D., during the year 2018 in the broad area of data mining. She has presented papers at conferences and published research articles in the field of data mining. Her areas of interest are Big Data, Data Mining and Software Metrics.

A. Umamakeswari Dean, School of Computing, SASTRA Deemed to be is an eminent researcher in the fields of Security in Wireless Sensor Network, Cloud computing, embedded systems and Internet of Things and has published many research papers over the last years in refereed and reputed journals and conferences.

Chapter 5
Engineering Full Stack IoT Systems with Distributed Processing Architecture—Software Engineering Challenges, Architectures and Tools

S. Thiruchadai Pandeeswari, S. Padmavathi and N. Hemamalini

Abstract New Age applications that stem from IoT, 5G systems, stream analytics, real time analytics, Industry4.0 have changed the way software are designed, developed, tested and maintained. These modern applications are highly scalable, highly dynamic, involve mobility and require frequent upgrades. Traditional computing paradigms and monolithic application architecture has become inefficient for engineering these applications which demand reusability, scalability and flexibility to implement upgrades. As a result, both computing paradigms and software engineering approaches have undergone major transition to suit the dynamic requirements of these applications. These applications may be engineered as highly cohesive, loosely coupled microservices that are easy to deploy and maintain. These microservices may also be deployed in such a way that processing happens in a distributed and independent manner. Concepts such as Fog computing and edge computing together with virtualization and Containerization provide support for this distributed processing architecture. This chapter aims to survey the existing contributions that have leveraged distributed computing paradigms and microservices architecture for the development of Full stack IoT applications. This chapter presents various challenges in developing IoT applications using microservices architecture and deploying them in a distributed processing environment such as Fog. This chapter intends to highlight various architectural patterns available for engineering microservices, challenges that are encountered while engineering large scale applications as microservices and number of tools and technologies that can be leveraged for the same.

S. Thiruchadai Pandeeswari (✉) · S. Padmavathi
Thiagarajar College of Engineering, Madurai, India
e-mail: eshwarimsp@tce.edu

S. Padmavathi
e-mail: spmcse@tce.edu

N. Hemamalini
Inspirisys Solutions Ltd., Chennai, India
e-mail: hemsush@gmail.com

© Springer Nature Switzerland AG 2020
J. Singh et al. (eds.), *A Journey Towards Bio-inspired Techniques in Software Engineering*,
Intelligent Systems Reference Library 185,
https://doi.org/10.1007/978-3-030-40928-9_5

71

5.1   Introduction

Modern applications of IoT such as Smart Surveillance systems, Smart health monitoring systems, Precision agriculture, Smart cities etc., are becoming more prevalent.
Engineering these applications pose greater challenges to traditional software development processes as use cases keep changing and added up every now and then. It is
quite challenging to estimate the scale of these applications as they scale up pretty
quickly. Upgrades need to be done quite fast and precise. It is very important to
note that these applications involve mobility, more number of service instantiations
and state of the art service provisioning. Hence these applications inherently require
a distributed processing architecture moving away from the centralized traditional
monolithic architecture. Monolithic architectural patterns are becoming ineffective to
engineer these new age applications for variety of reasons which has been highlighted
above. This chapter discusses various requirements of modern day IoT applications,
various computing paradigms available to realize these applications, changes in software engineering processes and guidelines that must be carried out while engineering
these applications.

5.2   Full Stack Development of IoT Applications

The term full stack development in connection with traditional applications refers to
the development of front end (user interaction), backend (processing, manipulating
and storing data) and databases. However, the full stack of IoT applications differ
to greater extent from the traditional web based applications. As mentioned in [1],
any IoT application would include the following layers: Perception layer, Transport
layer, Processing Layer, Application Layer, Business layer. The layered architecture
of IoT systems are shown in Fig. 5.1.

The perception layer contains sensors and end user devices which senses the physical parameters and produces data. All the data produced by the perception layer must
be further processed to produce useful information and decisions. Since the amount
of data produced by the sensors is usually huge and processing such voluminous
amount of data requires extensive resources and intensive processing capability, IoT
systems have Cloud as their integral part. The processing of huge amount of data
usually takes place at the cloud. Hence the processing layer is realized by integrating
cloud with the application. Transport layer is concerned with the transport of the
data produced at the perception layer to the processing layer i.e. cloud for processing. Application layer represents the services provided to the user in a consumable
form and associated logic. Business layer is concerned with business model of the
application. Hence the development of full stack IoT applications involves sensor
technology, communication protocols and middleware technologies as well.

Fig. 5.1 Layered
architecture of IoT systems

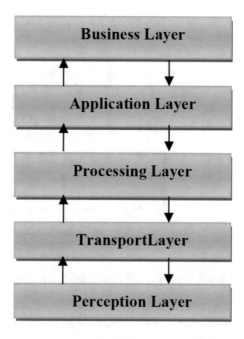

5.3   Requirements and Challenges of IoT Systems

Realizing full stack IoT applications such as automated cars, Smart Health monitoring systems and smart surveillance systems has their own challenges as discussed in [2]. These applications deal with real time data which needs to be processed quickly as the decisions are to be taken quickly in real time as well. This requires huge network bandwidth and associated latency needs to be limited. Also, it must be noted that the end user devices present in the perception layer are generally hand held and resource constrained. Hence there exists a stringent limitation on processing which the hand held devices can do. It is noteworthy that unlike traditional enterprise and web applications, the number of nodes involved in an application is very high and involve a fair degree of mobility. So it is quite challenging to make the security and software updates available to all the nodes, given that only a negligible-to-no application downtime is allowed. The challenges mentioned above necessitated a transition from centralized architecture with cloud at its center to a more distributed architecture for real time IoT systems. Many computing paradigms proposed as in [3–7] have been evolved to support this Geo-distributed processing architecture. Cloudlets and Fog play an important role in this transition to geo distributed processing. These paradigms involve an intermediate processing node between the actual processing layer and the perception layer i.e. cloud and data generation nodes. These intermediate processing nodes have decent processing capabilities and are closer to the edge i.e. data sources. Illustration of layered architecture of the IoT systems with intermediate distributed processing nodes is given in Fig. 5.2. Having intermediate

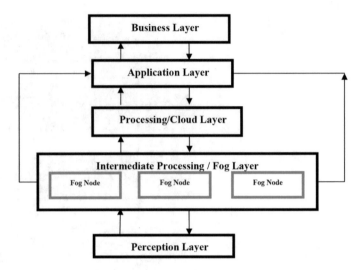

Fig. 5.2 Layered architecture of IoT systems with distributed intermediate processing nodes

processing nodes of relatively lower processing capability closer to the edge adds the following two major benefits:

- Quick processing of time critical tasks with reduced latency and accuracy.
- Eliminating the requirement of prohibitive amount of network bandwidth to transport all the data produced to the cloud. Only the resource critical tasks are carried forward to the cloud for further processing.

However, from [8] it is clear that addition of this intermediate processing layer comes with its own set of challenges in terms of service provisioning, monitoring, metering and ensuring quality of user experience. Provisioning time critical services in the intermediate nodes may require multiple service instantiations. It must also be noted that all the intermediate nodes may not necessarily be configured with same set of services.

5.4 Software Engineering Challenges of Developing Full Stack IoT Applications

From the above section that explained the challenges and requirements of Full stack IOT application development, it becomes clear that traditional software engineering approaches must be improved to suit the requirements. Traditional monolithic architecture supports all application functionalities as one full unit and focus on single large application deployment backed by an entire team of developers rather than multiple increments of deployments developed by a handful of developers. Though monolithic architecture has proved successful over a long time, the advent of IoT applications pose serious challenges to the monolithic architecture. Development of full stack IoT applications include implementation of number of functionalities that address the following issues on top of the applications original functionalities, say producing decisions from sensor data.

- Resource management among end user nodes, intermediate processing nodes and cloud—Based on the intensity and criticality of the tasks at hand, the tasks are scheduled to a suitable processing node. In order to do this, the tasks computational complexity is considered and state of the art resource allocation algorithms are used.
- Configuring intermediate processing nodes with services based on context—As the end user nodes involve mobility, they must be able to discover and receive services as they move away from the area of coverage of one processing node to another. In order to ensure this, service instantiations must be carried out at multiple nodes and the state of a service consumed by an end user node must be tracked and synchronized.
- Service usage monitoring and billing—as the consumer continue to get services from various intermediate processing nodes, their usage must be monitored, metered and billed.
- User participation monitoring—In some state of the art systems, the users also take part in building the processing and networking infrastructure. In such cases, the users are rewarded with incentives. So it becomes essential to monitor the user participation as well.
- Enhancing Quality of experience (QoE) by ensuring service continuity–For applications such as automated cars and intelligent transport systems, ensuring service continuity and Quality of Experience by end users is paramount as the end user nodes are mobile. This is accomplished by providing smooth handovers as user nodes move from the area of coverage of one processing node to another.

Hence a team that engineers full stack IoT application with distributed processing architecture must take care of the above mentioned points in addition to the original functional requirements of the application under development. It must also be noted that the modules of the application that implement the above functionalities are subjective to frequent changes as the applications scale up and down. Hence it is really difficult to tie all these requirements into one package and develop using monolithic architecture.

5.5 Microservices Architectural Patterns for Distributed IoT System

Microservices Architectural pattern considers the functionalities one at a time. Instead of packing all the functional requirements as one unit and developing them together, microservices architectural pattern aims at developing highly cohesive, loosely coupled services that are independent from one another. Any dependencies among the services are identified upfront and the services are piped accordingly. However it is very difficult to decompose all the functional requirements into independent functionalities. Thus it has been observed that it is best practice to separate less critical and loosely coupled functionalities and engineer them using microservices architectural pattern and engineer rest of the highly coupled functionalities using the traditional monolithic architecture.

Microservices can be engineered using number of architectural patterns as highlighted in [9]. Some of the familiar and suitable architectural patterns of microservices are discussed in connection with developing IoT applications below.

5.5.1 Scale Cube

This pattern defined [10] uses a three dimensional approach for scaling the microservices. This scale cube pattern provides a guideline for decomposing functional requirements from a monolith applications and engineering them with microservices. According to scale cube, the functional requirements are decomposed into smaller, independent and deployable microservices and are mapped on to the Y-axis of the cube. The multiple instantiations of the same service at geo distributed locations are mapped on to the X-axis. The shared slices of databases are mapped on to the Z-axis of the cube. The slices of databases in the Z-axis represent the subset of database that would be hit by a specific microservice. Thus, scale cube provides guidelines when the functionalities are split from the monolith application to be realized as microservices. The following diagram given in [10] explains decomposition of monolith application and mapping the decomposed functionalities as microservices. Ideally the functions must be decomposed in such a way that they perform a single functionality that cannot be further decomposed in order to be realized as a microservice. Scale cube proves to be suitable architecture for developing a full stack IoT application as it recognizes multiple instantiation of the same service. X-axis scaling needs to be carried out for provisioning services in multiple cloudlets or fog nodes. Similarly decomposing the functionalities into autonomously deployable services i.e. Y-axis scaling may be used to realize functionalities that are less critical and require frequent upgrades. Z-axis scaling i.e. sharding databases becomes inevitable with X-axis and Y-axis scaling as each instance of service runs a subset of database (Fig. 5.3).

5.5.2 Event—Driven Architecture

Scale cube have Z-axis scaling where each instantiation of a service affects a specific subset of database. Normally services use message brokers to update state information to other services that need to be mutually aware of each other. However, this

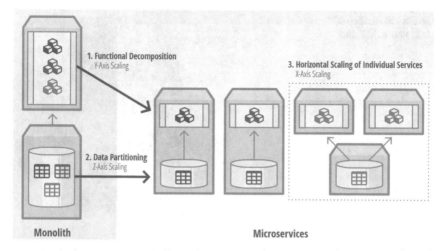

Fig. 5.3 Transition from monolith to microservices architecture based on scale cube

tends to affect the consistency of the databases. Event driven architecture proposed in [11] helps to address this. When microservices are engineered based on event driven architecture, each service publish an event whenever it updates its data. Other services that depend on the given service must be subscribed to the event. This enables data consistency.

For example, whenever a service given to a user is handed over from one processing node to another, an event "Service Handover" is published with the details of the service. The neighboring nodes which are present near may be subscribed to the event so that when they see a Service handover event getting published, it gets ready for next set of actions. Publishing an event after completion of a transaction by a service helps to pipe services whose execution depends on each other's state.

5.5.3   Hexagonal Architecture

Hexagonal architecture, also known as "Ports and adapters" architecture decouples the application's core logic and keeps the core logic abstracted from the non critical services through ports and adapters. This architecture was originally proposed in [12]. The core logic is engineered with necessary APIs called ports through which the services connect with the core logic. The interface between the service and core logic of the application is known as adapters. This reduces the complexity and increases decoupling. In the Fig. 5.4 referred from [12], the yellow colored portion represents core application logic. The red colored portion represents ports i.e. APIs using which the auxiliary services connect with the core logic. The blue structures represent the auxiliary services that can connect to the core logic using an inbuilt interface and the ports in the core logic. In case of large scale distributed IoT applications, it is not

Fig. 5.4 Illustration of
hexagonal architecture

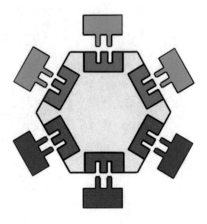

possible to realize the entire application using microservices architecture. Decomposition of all the functionalities in adherence to Single responsibility principle is not realistic. Hence hexagonal architecture proves suitable for full stack IoT application development. The important and core functionalities of the application could be retained as highly coupled unit of functions. This highly coupled core unit must be engineered with appropriate APIs for sending inputs to and receiving outputs from auxiliary services. Auxiliary functionalities such as service metering, billing and user participation monitoring could be realized as services with appropriate interfaces to connect with the core application logic using its APIs. The diagram illustrates hexagonal architecture for developing applications.

Thus scale cube, Hexagonal and Event Sourcing are few of the many software architectural patterns that suit the development of full stack IoT applications combining both the monolith and microservices architecture.

5.6   Tools for Microservices Implementation

Implementation of Microservices architecture requires the services to be orchestrated properly. Services Orchestration has the following important aspects.

- Service discovery—which requires the available services to be published in a common registry so that the services become mutually aware of each other.
- Service deployment—follows the phase of service development. In an efficient smart IoT system, Services must be able to self configure and deploy them autonomously based on context. Necessary infrastructure must be made available for this desired autonomous service deployment.
- Service Monitoring—the deployed services must be monitored as they get consumed. This is paramount in terms of service metering and billing. This section highlights various tools and technologies that can be leveraged for the development of microservices.

- Meta-information Maintenance—Since the Microservices are deployed and provisioned autonomously, there is an important need to maintain the meta information about the state of these services centrally with consistencies.

Implementation of these aspects of the microservices requires specific tools and technologies. This section highlights various tools and technologies available for the implementation of microservices implementation.

5.6.1 Docker Containers for Service Deployment

Docker containerization has enormously enhanced the ease and efficiency of microservices implementation. Docker containers provide a light weight virtualized environment for service deployment. Docker containers encapsulate all the necessary dependencies for executing a service into one portable light weight container. Docker application is similar to virtualization but however differ slightly from the virtualization in terms of OS virtualization and image building. For example, if a service requires deployment of Tomcat server then Docker makes it simple by the following steps

- Configure a single Docker image with Tomcat server
- Publish the image to a registry such Docker hub
- Instantiate a Docker Container by adding required images from the Docker hub

A Docker image needs to be configured only once and can be instantiated into any number of containers. Instantiation of container is also straightforward with an appropriate Docker file. This makes service deployment pretty simple rather than configuring Tomcat server manually into the distributed intermediate processing nodes where the services are to be deployed. Docker containers can also be easily ported. The booting up time of a container is very less when compared to that of a virtual machine. So Docker Containers have proved to be a very effective technology for configuring services into processing nodes especially for IoT systems with distributed processing model where latency is a very crucial parameter.

5.6.2 Tools for Service Provisioning

As mentioned earlier, Service discovery is an important aspect of microservices architecture implementation. Some of the tools for Service Discovery include Apache Zookeeper, Consul and Apache Storm.

Apache Zookeeper is an open source tool from Apache foundation that provides centralized services to distributed applications. From [13], the services offered by Apache Zookeeper for synchronization and co-ordination of distributed services can be observed. It also offers services for naming, configuration management and other group services.

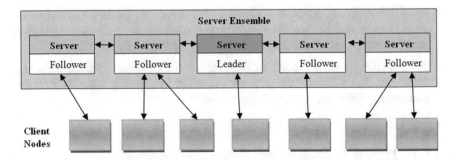

Fig. 5.5 Working architecture of Apache Zookeeper

Zookeeper maintains an ensemble of servers and chooses a leader among them. Majority of the servers of the ensemble must be up and running for a client to get service from one of the servers. The configuration information are stored among the ensemble of servers in a distributed manner. Any change in the configuration will be committed only when the commit pass through majority of the servers. A client node may get connected to any one of the servers of the ensemble and start getting the services and making changes. Until majority of the servers are up and running, the commits made by the client nodes will be updated in the distributed databases. In case of conflicts, the changes made by leader node get committed.

The Fig. 5.5 shows the architecture of Apache ZooKeeper. Thus Apache Zookeeper is an excellent tool for provisioning distributed services that takes care of the typical problems of distributed applications such as deadlocks, conflicts and synchronization. Leveraging zookeeper for microservices service provisioning in the context of IoT systems would be ideal. This facilitates the software developer to concentrate on the application's logic development rather than working on the coordination of microservices. The microservices when provisioned through the zookeeper's ensemble of servers, allows client nodes to easily connect to one of the services and start making transactions. The changes made by a particular node while executing a microservice would be committed to all the servers; thereby it avoids conflicts, deadlocks and inconsistencies. **Consul** is a software tool that helps to coordinate and maintain multiple services. It is offered by Hashicorp. As given in [13], Consul is configured as a cluster which contains number of networked nodes that run registered services. The networked nodes may simply run a consul agent or act as consul servers. When a network node runs consul agent, it helps the network node to discover and consume services. The consul servers are where the services are actually registered and any changes in service provisioning could be carried out only by the consul servers. The Consul servers provide a centralized naming registry that helps client nodes that run consul agents to discover services. Consul has a component called 'Consul Connect' that makes use of Service graphs to implement the inter service communication policies. The service graphs together with TLS are used to ensure service security. Services basically discover other services and talk to them based on service

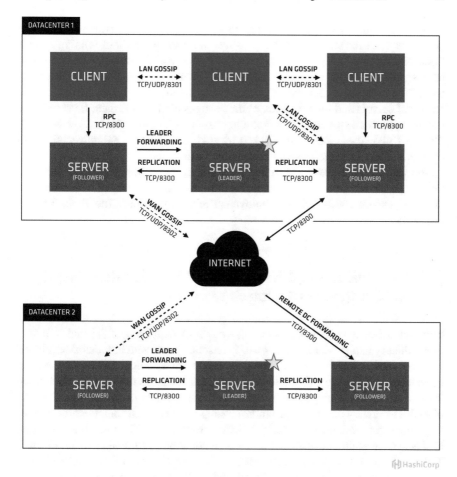

Fig. 5.6 High level architecture consul framework

graphs. Service Registrations and removal happen through HTTP API. Consul has another component called Key Value store that makes service configurations simple. Instead of maintaining configuration information in pieces across multiple servers, the key value store pushes whole of the configuration files into all the network nodes so that the nodes take configuration information by dialing into the loopback address itself. Consul makes use of gossip protocol for communicating state information among consul agents i.e. it doesn't maintain a centralized repository for maintaining state information of services to which the agents communicate to, when they need information. Event based updates are passed on among the consul agents using gossip dissemination protocols. To manage conflicts, Consul servers elect a leader using Raft algorithm which is consensus based algorithm. The architecture is Consul is given in the Fig. 5.6 referred from [13]. **Etcd** is yet another open source offering for provisioning consistent distributed applications. It is managed by Cloud Native computing Foundation and it is available on github [14]. It provides a consistent key value store for storing data that belongs to many instances of running service instances. It is also a Raft based system that elects leader for cluster. All the client requests that require consensus among cluster members are handled by leader and rest of the requests are handled by other cluster members. It saves the data across cluster of servers and reliably updates the changes made to the data. It uses simple http and JSON constructs for its communications with the services. It is noteworthy that Kubernetes clusters are backed by etcd. The cluster state information and configuration of kubernetes services are stored using etcd.

Above mentioned are few tools that are available for orchestrating microservices. Other than the mentioned tools, whole lot of tools and frameworks are available for the development, discovery and deployment of microservices. The Table 5.1 lists few of them.

5.7 Case Study: Smart Nutrition Monitoring System Based on Microservices Architecture

Consider a Smart Nutrition monitoring system as explained in [7] that monitors users' dietary habits with the help of IoT sensors. The system has a kiosk equipped with IoT sensors that collect data about the food such as weight and volume. All the users have to do is, deposit the food in the kiosk for some time before consuming it so that the kiosk equipped with sensors collect dietary information from the food placed. The dietary data are then analyzed using number of data analytics techniques to produce recommendations and predictions to the user. The overall architecture of this system without distributed processing could be illustrated as in the Fig. 5.7. The high level architecture presented shows end user devices such as mobile phones, wearable gadgets and Kiosks equipped with IoT sensors in the perception layer. These devices present in the perception layer collect user dietary data and pass on to the cloud servers. Cloud servers have databases for storing the dietary data from the users. Cloud servers also house core application logic for applying analytics and visualization over the user dietary data. The application layer contains the user interfaces for the web and mobile application. This smart nutrition monitoring system may be further made efficient by introducing intermediate processing nodes i.e. fog nodes in addition to centralized cloud, so that part of the processing could take place in the fog nodes. An enhanced architecture with intermediate fog nodes is shown in Fig. 5.8. The enhanced fog based architecture includes a special layer called Fog layer that includes fog nodes. Anything from a physical server to base station can be configured as a fog node. The enhanced fog based smart nutrition monitoring system necessitates transition from monolithic architecture in which the cloud servers house the entire application logic and databases so that the services are provisioned centrally. Since the fog nodes come with the capability of decent processing, the fog nodes can be leveraged for service provisioning. In that case, the core application

Table 5.1 Tools and frameworks available for microservices implementation

Name of the tool/framework	Usage	Github repository
RestExpress	Java framework for creating scalable, containerless, RESTful microservices. Uses Netty for network communications, JSON for message passing and metrics plugin for health checks	https://github.com/RestExpress/RestExpress
Spring Boot	Java based framework that helps creating microservices, get them up and running quickly with minimal configurations. Spring cloud which is built on top of Spring boot provides features to quickly develop and deploy microservices with embedded servers	https://github.com/spring-cloud
Eureka	Eureka is REST based discovery service adopted by AWS for locating services. Spring boot also leverages Eureka registry Service	https://github.com/spring-cloud-samples/eureka
Apache ActiveMQ	Offered by Apache foundation, it is a popular multi-protocol java based messaging server. Supports popular messaging protocols such as AMQP, STOMP and MQTT	https://github.com/apache/activemq-apollo
ScaleCube	Provides low latency reactive microservices library for peer-to-peer service registry and discovery based on gossip protocol, without single point-of-failure or bottlenecks	https://github.com/scalecube/scalecube-services
Jersey	Framework for developing RESTful web services. Provides its own API for extending JAX-RS toolkit	https://github.com/jersey
GO Kit	Toolkit in GO language for writing microservices with standard libraries	https://github.com/go-kit/kit
Apache Storm	Open source distributed real time computation tool. Used to carry out real time processing of streams of data in a distributed manner	https://github.com/apache/storm
Apache Kafka	Distributed Streaming Platform. Helps developing real time scalable stream processing applications	https://github.com/apache/kafka

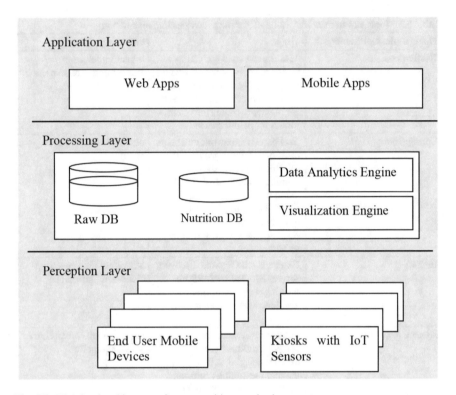

Fig. 5.7 High level architecture of smart nutrition monitoring system

logic needs to be broken into number of autonomous logics that can be developed and deployed as independent services. The fog layer contains number of fog nodes. Each fog node may be configured with different services. In Fig. 5.8, for illustration purposes, three fog nodes are shown. The colored blocks inside each fog node represent different services deployed in the fog nodes. Functionalities that are chosen to be architected as microservice must be highly cohesive. State dependence on other functionalities must be strictly eliminated. Four services chosen for illustration in the Fig. 5.8 are as follows:

- Data Preprocessing Service—This includes Data cleaning and trimming. Each fog node cleanses the data sent by the perception nodes in their area of coverage.
- Data Visualization service—This represents simple visualization of non persistent data.
- High calorie Food alert service—This service is concerned with calculating the calories of the food placed in kiosk and alerting the user if the total calorie exceeds the total upper limit for a single serving.
- Expert Recommendation service—This service is concerned with setting up adhoc sessions between users and experts.

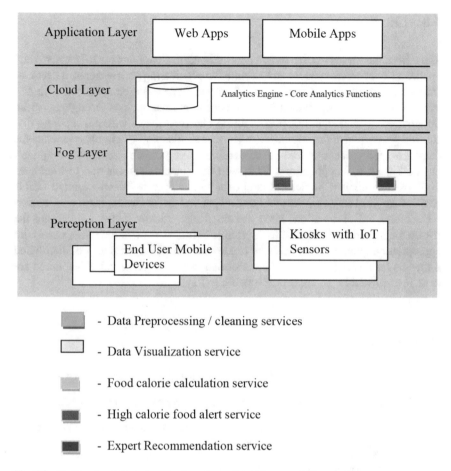

Fig. 5.8 Highlevel architecture of fog based smart nutrition monitoring system

The above functionalities are all state independent of other functionalities and thus can be realized using microservices. Also, they satisfy the single responsibility principle of microservices architecture, where each microservice is associated with one autonomous functionality. Thus breaking up these functionalities from the core logic simplifies the core logic and prevents the application from centrally dependent on cloud. Also, when many auxiliary functions are realized as autonomous microservices that can be independently deployed, provisioned and monitored the application become easily scalable and more flexible.

5.8   Conclusion

Thus, from the survey carried out on new computing paradigms for IoT systems and applicability of Microservices architecture for engineering such systems, it becomes clear that performance of Full stack IoT applications could be enhanced with distributed processing. Significant improvement in metrics such as throughput, mean time between failures and Capex could be observed in the works leveraging distributed processing architecture. Microservices architecture inherently complements the distributed processing and hence considered very much suitable for engineering Full stack IoT applications. Rich set of architectures, tools and technologies available for efficient implementation of microservices has been explored in this chapter. Applicability of these architectural patterns in the development of microservices based IoT systems has been discussed theoretically while highlighting the requirements and challenges of IoT systems. In future, verification of the same with appropriate implementation could be carried out. Also, the efficiency of distributed protocols for message passing, state information sharing and conflict resolution are to be investigated in the context of modern IoT systems.

References

1. Sethi, P., Sarangi, S.R.: Internet of things: architectures, protocols, and applications. J. Electr. Comput. Eng. (2017)
2. Chiang, M., Zhang, T.: Fog and IoT: an overview of research opportunities. IEEE Internet Things J. **3**(6), 854–864 (2016)
3. Satyanarayanan, M., Bahl, V., Caceres, R., Davies, N.: The case for vm-based cloudlets in mobile computing. IEEE Pervasive Comput. (2009)
4. Van Eyk, E., Toader, L., Talluri, S., Versluis, L., Uță, A., Iosup, A.: Serverless is more: from paas to present cloud computing. IEEE Internet Comput. **22**(5), 8–17 (2018)
5. Nastic, S., Rausch, T., Scekic, O., Dustdar, S., Gusev, M., Koteska, B., Prodan, R.: A serverless real-time data analytics platform for edge computing. IEEE Internet Comput. **21**(4), 64–71 (2018)
6. Donassolo, B., Fajjari, I., Legrand, A., Mertikopoulos, P.: Fog based framework for IoT service provisioning. In: IEEE consumer communications and networking conference (2018)
7. Mehdipour, F., Javadi, B., Mahanti, A., Ramirez-Prado, G.: Fog computing realization for big data analytics. In: Fog and Edge Computing: Principles and Paradigms, pp. 259–290 (2019)
8. Pandeeswari, S.T., Padmavathi, S.: Fog based architectures for IoT: survey on motivations, challenges and solution perspectives. In: Auer, M., Ram, B.K. (eds.) Cyber-Physical Systems and Digital Twins. REV2019 2019. Lecture Notes in Networks and Systems, vol. 80. Springer, Cham (2020)
9. Williams, A., et al.: Applications and Microservices with Dockers and Containers, vol. 2. The New Stack
10. Abbott, M.L., Fisher, M.T.: The art of scalability: scalable web architecture, processes, and organizations for the modern enterprise. Pearson Education, London (2009)
11. Michelson, B.M.: Event-driven architecture overview. Patricia Seybold Group 2(12) (2006)

12. Cockburn, A.: Hexagonal Architecture: Ports and Adapters ("Object Structural") (2008)
13. Haloi, S.: Apache Zookeeper Essentials. Packt Publishing Ltd, Birmingham (2015)
14. Consul architecture, https://www.consul.io/docs/internals/architecture.html
15. Webb, P., Syer, D., Long, J., Nicoll, S., Winch, R., Wilkinson, A.: Spring boot reference guide, p. 24. Part IV, Spring Boot features (2013)
16. Etcd github repository, https://github.com/etcd-io/etcd

Chapter 6
Chaos-Based Modified Morphological Genetic Algorithm for Effort Estimation in Agile Software Development

Saurabh Bilgaiyan, Prabin Kumar Panigrahi and Samaresh Mishra

Abstract One of the most critical and important aspects of any software development project is the estimation of cost and effort, as the success or failure of the entire project is largely dependent on the accuracy of these estimations. For any software development project, several methods such as waterfall, prototyping etc. exist, but the agile methods have prevailed in terms of its efficiency and implementation in solving the problems related to the projects thus substituting the traditional methodologies. Agile methods have become much popular recently because of its ability to adopt to the changing dynamics (requirements) of software projects. This dynamic nature makes the task of estimation even more challenging than the traditional methodologies present. Thus, it becomes convenient to accurately estimate the effort and cost while adopting the agile methods, for which various techniques have already been proposed such as analogy, dis-aggregation, expert opinion etc, but none among the same have a proper mathematical model. This work has presented a novice method from the domain of evolutionary algorithms. The work is based on mathematical morphology (MM) consisting of a hybrid-artificial neuron (Dilation-Erosion perceptron (DEP)) extended from the concept of complete lattice theory (CLT). Authors have presented a chaotically modified genetic algorithm (CMGA) to build the DEP-CMGA model for solving the software development effort estimation (SDEE) problem. Calibration of the proposed model was done using data collected from 21 software projects based on agile software development (ASD). Four different statistics were used for determining the precision of the model and the results were compared with the one's obtained using the best available model in literature.

S. Bilgaiyan (✉) · S. Mishra
School of Computer Engineering, Kalinga Institute of Industrial Technology, Deemed to be
University, Bhubaneswar, India
e-mail: saurabhbilgaiyan01@gmail.com

S. Mishra
e-mail: samaresh2@gmail.com

P. K. Panigrahi
C. V. Raman College of Engineering, Bhubaneswar, India
e-mail: prabinprakash1@gmail.com

© Springer Nature Switzerland AG 2020
J. Singh et al. (eds.), *A Journey Towards Bio-inspired Techniques in Software Engineering*,
Intelligent Systems Reference Library 185,
https://doi.org/10.1007/978-3-030-40928-9_6

Keywords Evolutionary algorithms · SDEE · Complete lattice theory ·
Mathematical morphology · Agile software development

6.1 Introduction

In the past few years, a phase shift was observed from traditional methodologies
to more sophisticated agile methods of software development. According to latest
survey, the percentage of industries that have never applied agile methods is as low
as 3.4% [1]. Agile methodologies basically incorporate a number of methods sharing
same goals and values in-order to achieve collaborative software development based
on the concepts of multiple iterations, incremental improvement, feedback sharing,
documentation and specification [2, 3]. Traditionally agile methods were believed
to be best suited for small correlated teams, but its efficiency in small teams inspired
its use in large scale and very large-scale software development project [4].

Since the articulation of agile manifesto back in 2001, they have developed and
delivered software products with satisfaction and under the given time-frame thus
bringing about unprecedented changes to the software filed [5].

Industrial adoption of several manifestation such as scrum, crystal and eXtreme
programming has made agile software development a mainstream in the software
development industry [6]. A major cause (as identified by Standish group [7]) of
software failure is incorrect and incomplete requirements. Requirement prioritization
is the responsibility of the client. Thus, the metrics must be able to calculate the
project's success rate accurately.

Project cost, effort and quality are the determining factors for the success of the
project. Using this metric, the project success rate can be determined even before
the starting of the actual project. The traditional prediction models are either too
sophisticated to implement or rather unreliable, thus presenting a problem in the
agile development [1].

Morphological neural networks have been extensively used with many diversified
applications in literature. Basically, it's a type of artificial neural network with MM
frameworks having algebraic foundations in complete lattice theory (CLT) [8]. In
MNNs' an elementary MM operation is performed at each node of the network.
In classical neural networks, merging and processing of information takes place by
multiplying and summing the output values with corresponding weights whereas in
MNN process, summation with corresponding weights is followed by selection of
highest value from that [8, 9].

The non-differentiability problem which affects the dilation and erosion operators
is a major drawback of the classical DEP model and to overcome this problem an
efficient methodology is needed [8].

This paper presents an evolutionary morphological approach for solving the
SDEE problem. In the proposed model, DEP with a modified genetic algorithm
is used, termed as CMGA (DEP). Data from 21 different software projects was used
for testing the proposed model and three performance metrics were used for perfor-

mance evaluation: mean magnitude of relative error (MMRE), prediction (PRED) and evaluation function (EF) [8, 11, 12].

The organization of this paper is as follows: Sect. 6.2 contains the comparable related work followed by DEP and MNN fundamentals in Sect. 6.3. Section 6.4 focuses on the proposed swarm intelligence based learning method, Sect. 6.5 focuses on the performance metrics followed by simulations and experimental analysis in Sect. 6.6. Section 6.7 contains the conclusion and future work.

6.2 Related Work

Soft computing techniques has been widely applied in the various fields such as machine learning, AI, data preprocessing, etc. It involves processes that involve indirect, approximate solutions instead of binary algorithms, widely considered to include such technologies as genetic algorithm (GA), fuzzy logic and neural networks. A genetic algorithm is a heuristic technique that attempts to find high-quality solutions to large and complex optimization problem [13].

Literature shows that agile methods is better suited for software project developments as compared to other models such as Feature Driven Development (FDD), Extreme Programming (XP) etc. basically because of its dynamic nature [14].

MNNs have been used in many diversified applications throughout the literature and also the techniques of DEP and soft computing have been successfully implemented in many MNNs for solving the prediction problems which makes use of MM operators on CLT [8, 9].

Following are some related works found in literature regarding the methods discussed above:

Araujo et al. [15] used hybrid methodologies for designing a MRL perceptron in order to solve the SDCE problem, where the perceptron's parameters were optimized using MGA. Braga et al. [16] improved the accuracy of estimated software effort by using ML methods with robust confidence interval. This method is independent of training set's probability distribution of error. Oliveira et al. [17] presented a ML based feature selection and parameter optimization method for SDEE based on GA whose accuracy of prediction was improved using input feature selection and parameter optimization. Araujo et al. [18] presented a MRL approach to solve the SDEE problem. The proposed approach comprised of a hybrid morphological model having linear combination of a linear finite impulse response (FIR) operator termed as MRL filter and a nonlinear morphological rank (MR) operator. The parameters are adjusted using a gradient steepest descent (GSD) method. Araujo et al. [19] proposed a morphological approach by using gradient method based on MM having algebraic foundations in lattice theory. This model used a combination of dilation and erosion operators from the domain of MM. Training of DEP done by GSD method using back propagation (BP) algorithm. Araujo et al. [8] proposed a DEP which is basically a hybrid artificial neuron which uses (MGA) for setting up its parameters and it was successfully tested for solving SDEE problem. Araujo et al. [20] used least mean square algorithm and MGA to design a MRL perceptron called as morphological rank linear hybrid intelligent design for solving SDCE problem. Araujo et al. [21] proposed a shift invariant based morphological system for solving SDCE problem consisting of hybrid morphological model and it uses FIR and MR operators. Srdjana et al. [1] proposed a Bayesian network model suitable for effort prediction in agile methods. Implementation of this model can be done in the planning stage itself. A single software company is used for elicitation of data using complete agile projects. Results indicated good prediction accuracy. Aditi panda et al. [12] presented a story point approach for effort calculation of agile software projects. Various neural networks such as probabilistic neural network (PNN), general regression neural network (GRNN), etc were used. The performance generated hence were analyzed. Jitender Choudhari et al. [22] proposed a software maintenance effort estimation model (SMEEM) for the estimation of software maintenance. The model used SPA for calculation of the maintenance volume and adjustment factors affecting the effort estimations. This model was suitable for agile and eXtreme programming environments. Ziauddin et al. [11] proposed an effort estimation model for solving SDEE problem in ASD projects. Calibration of the model was done with the help of empirical data collected from 21 software projects.

6.3 CMGA Process

6.3.1 Basic Principles of MNN

MNN's are basically artificial neural networks (ANN) performing elementary operations from MM at each network processing unit, between complete lattices [8]. This proposed work is mainly focuses on the lattice algebraic characterization of MNN.

Lattices are well-defined as a partially ordered set L whose every single finite nonempty subset has its infimum and supremum in lattice L. For any $A \subseteq L$ infimum of A is represented by the symbol $\wedge_n A$. If $A = \{a^n, n \in N\}$ for an index set N, it is common to use $\wedge_{n \in N} x^n$ instead $\wedge_n A$. Supremum of A is also defined in a similar manner [8, 9].

For any two given lattices L_1 and L_2, a mapping $\psi : L_1 \rightarrow L_2$ is increasing if the following condition is met for all a, b: [8]

$$a \leq b \Rightarrow \psi(a) \leq \psi(b) \tag{6.1}$$

For a given lattice L, a partial ordering on L^k is defined by:

$$(a_1, a_2, a_3 \ldots a_k) \leq (b_1, b_2, b_3 \ldots b_k) \Leftrightarrow a_m \leq b_m, m = 1, 2, 3 \ldots k \tag{6.2}$$

Here, L^k which is the resultant partially ordered set also called product lattice [8, 9]. If L has an infimum and supremum for every single finite or infinite, non-empty subset, then the lattice L is said to be a complete lattice (CL). Also by following this the product lattice Lk is also a CL L^k [10].

CLs' are considered as an appropriate theoretical framework for MMs. A major issue here is the decomposition of mappings between CLs in terms of elementary morphological operators [8].

Several theorems have been proposed by Bannon and Barrera [10] for decomposition of mappings among CLs in terms of basic MM operations. Suppose that algebraic erosion and algebraic dilation δ be operators from CL L_1 to CL L_2 and $Y \subseteq L$, then

$$\varepsilon(\wedge y) = \wedge_{y \in Y} \varepsilon(y) \tag{6.3}$$

$$\delta(\vee y) = \vee_{y \in Y} \delta(y) \tag{6.4}$$

Bannon and Barrera [10] proposed that between two complete lattices L_1 & L_2, an increasing mapping $\psi : L_1 \to L_2$ can be shown by the decomposition given as: infimum of dilation operators and supremum of erosion operators. For given indices P and Q, there are dilation and erosion operators represented by δ^p and ε^q respectively given by:

$$\psi = \wedge_{p \in P} \delta^p \tag{6.5}$$

$$\psi = \vee_{q \in Q} \varepsilon^q \tag{6.6}$$

Learning algorithms of several MNNs have their basis in these decomposition theorems. The elementary operators from MM in this decomposition adopt a special form requiring algebraic structures beside the CL structure. Here, the main focus is on the complete lattice $\mathfrak{R}_{\pm\infty}$ as the function $\mathfrak{R}_{\pm\infty}^h \to \mathfrak{R}_{\pm\infty}$ (h being the number of variables in the software project) can be used to model the SDEE problems [8, 10, 23].

Here, $\mathfrak{R}$ is the set of finite elements of $\mathfrak{R}_{\pm}$ which builds a group having conventional addition. It is defined as lattice ordered group also called as l-o group and every group's translation using "+" is an isotone [8]. A lattice is L is called a l-o group if for every $a, b \in L$ and $a \leq b$ [9], "+" is the group operation and it has:

$$x + a + y \leq x + b + y \forall x, y \in L \tag{6.7}$$

The bounded lattice ordered group or bounded l-o group extension considered in the minimax theory states a bounded lattice D having a set of finite elements $E \subset F$ represented by $D\backslash\{+\infty, -\infty\}$ where, $+\infty$ denotes $\vee D$, greatest element of D, and $-\infty$ denotes $\wedge D$, the least element of D, which generates a group [9, 23].

This work focuses on a real number set $\mathfrak{R}_{\pm}$ which is a complete l-o group extension. Also the group operation "+" can be extended [9]:

$$x + (+\infty) = +\infty + x = +\infty, \forall x \in \mathfrak{R} \cup \{+\infty\} \tag{6.8}$$

$$x + (-\infty) = -\infty + x = -\infty, \forall x \in \mathfrak{R} \cup \{-\infty\} \tag{6.9}$$

For the given matrices $A \in \mathfrak{R}^{t \times u}$ and $B \in \mathfrak{R}^{u \times v}$, the max product and min product is given by $C = A \vee B$ and $D = A \wedge B$ respectively, defined by:

$$c_{ij} = \vee_{k=1}^{u}(a_{ik} + b_{kj}) \tag{6.10}$$

$$d_{ij} = \wedge_{k=1}^{u}(a_{ik} + b_{kj}) \tag{6.11}$$

Hence, the operators of MM can be used to define a matrix product where δA and $\varepsilon A: \Re_{\pm\infty}^{h} \to \Re_{\pm\infty}^{g} \forall A \in \Re^{h \times g}$ are given as:

$$\varepsilon_A(x) = A^T \wedge x \tag{6.12}$$

$$\delta_A(x) = A^T \vee x \tag{6.13}$$

It is seen that operators δ_A and ε_A represent the algebraic dilation and erosion from $\Re_{\pm\infty}^{h}$ to $\Re_{\pm\infty}^{g}$ which are CL. So δ_A and ε_A represents dilation and erosion of every form [8, 9]. The above statement along with Eqs. 6.5 and 6.6 can be used to approximate the increasing function $\psi : \Re^h \to \Re$ in terms of vectors t_p and $w_q \in \Re^h$ and indices $\bar{P}$ and $\bar{Q}$:

$$\psi \cong \vee_{p \in P} \varepsilon_{t^p} \tag{6.14}$$

$$\psi \cong \wedge_{q \in Q} \delta_{w^q} \tag{6.15}$$

Approximation in terms of the vectors $t, w \in \Re^h$ can be done for the increasing function mapping $\psi : \Re^h \to \Re$ for indices $\bar{P}$ and $\bar{Q} = 1$, given by:

$$\psi \cong \varepsilon_t \tag{6.16}$$

$$\psi \cong \delta_w \tag{6.17}$$

Hypothesis of Eqs. 6.16 and 6.17 form a basis for the SDEE problems using morphological perceptrons.

6.3.2 Basic Principles of DEP

DEP is a class of hybrid morphological perceptron which is basically a convex combination of dilation and erosion operators of MM on CLT, and is used to solve the SDEE problem [8, 9].

Let $w_1, w_2, w_3 \ldots w_n \in \Re^n$ be a real valued input signal in a n-point moving window. Assume v denote the outcome of the DEP. Then a shift invariant morphological system with $w \to v$ as a local signal transformation is used to define the DEP as:

$$v = \lambda\alpha + (1 - \lambda)\beta, \lambda \in [0, 1] \tag{6.18}$$

where,

$$\alpha = \delta_a(w) = \vee_{i=1}^{n}(w_i + a_i) \tag{6.19}$$

$$\beta = \varepsilon_b(w) = \wedge_{i=1}^{n}(w_i + b_i) \tag{6.20}$$

Here, $\lambda \in \Re$, $a, b \in \Re^n$ are the constructing elements (morphological operators) of DEP denoted by $a = a_1, a_2, a_3 \ldots a_n$ and $b = b_1, b_2, b_3 \ldots b_n$ with n being the dimensions of input signal. There exists an inverse proportionality relation among the elements of DEP such that when one element's contribution increases, the other one tends to decrease and vice-versa.

6.4 Proposed Evolutionary DEP(CMGA) Process

The weight vector in case of DEP(CMGA) learning procedure is defined by [24]:

$$w = (a, b, \lambda) \tag{6.21}$$

Here, a, b and λ represent the parameters required for the adjustment during the DEP(CMGA) learning process until the termination criteria is met. The adjustments are being performed according to the error criteria till the convergence of CMGA iterations [24].

Here, $w_j^{(g)}$ represent the candidate weight vector of the gth generation's jth individual. For assessing and adjusting the quality of the weight vector, a fitness function (FF) is used, given by [8, 24]:

$$ff(w_j^{(g)}) = \frac{1}{N} \sum_{k=1}^{N} e^2(k) \tag{6.22}$$

Here, N represents the count of input patterns and $e(k)$ is the count of instantaneous error (IE) represented by:

$$e(k) = d(k) - y(k) \tag{6.23}$$

Here, $d(k)$ represents the desired outcome signal and $y(k)$ represents the obtained outcome for a given sample k.

6.4.1 Population Initialization by Chaotic Opposition (PICO) Based Learning Method

In order to obtain better output from this process, a chaos based opposition method for population initialization was adopted instead of random initialization. The sensitivity dependence on initial conditions and randomness of chaos maps were the prime reason for selecting them in this process. They are also used for search space information extraction and for increasing the population diversity. The convergence speed of the CMGA algorithm also increases by adopting this method [25]. A sinusoidal iterator was chosen for this method given by:

$$cho_{t+1} = \sin(\pi cho_t), \, cho_t \in [0, 1], \, t = 0, 1, 2, 3 \ldots max_t \qquad (6.24)$$

Here, t is the iteration counter and maximum iterations is represented by max_t. The following steps are to be considered for PICO algorithm:

Algorithm 1: Steps for PICO Algorithm

---
begin
    (PICO) [25];
    Data: Set max chaotic iterations as $Max_t = 300$
    $Size_{pop} = U$
    Result: $Initial_{pop}$
    for $i=1$ to U **do**
        for $j=1$ to D **do**
            ▷ Here, D represent the vector dimention of a individual;
        Randomly initialize variable $cho_{0j} \in [0, 1]$;
        for $r=1$ to Max_t **do**
            | $cho_{rj} = \sin(\pi cho_{r-1,j})$
        end
        $x_{ij} = x_{min_j} + cho_{rj}(x_{max_j} - x_{min_j})$
        end
    end
    (—Start Opposition Based Method—);
    for $i=1$ to U **do**
        for $j=1$ to D **do**
        | $opx_{ij} = x_{minj} + x_{maxj} - x_{ij}$
        end
    end
    Select U number of fittest individuals from $(X(U) \cup OPX(U)$ and set
    them as initial population;
        ▷ Here, X(U) and OPX(U) represent the set of produced U individuals
end

6.4.2 Evolutionary Learning Process

A loop comprising of phases for minimizing the fitness function fit $\mathfrak{R}^h \to \mathfrak{R}$: is started by the CMGA process as defined in Eq. 6.22. Here the term h refers to the dimensionality of the weight vector of DEP model. The loop contains steps of selection, crossover and mutation operators. In selection, a parent pair of chromosome are selected, then operations of crossover and mutation operators are applied on the parents to generate new off-springs which generates the new population (population size taken here is M = 10). The process is repeated till the termination criteria is satisfied. The solution of the problem is the best individual selected form the population [24].

6.4.2.1   Crossover Process

For exchanging information between two parents, (vectors $w_a^{(g)}$ and $w_b^{(g)} \in \Re^h$, where a, b are indices within range [1, M]) the crossover operator is used where a roulette wheel approach is used for selecting the two parents. Crossover operators are used for the recombination process which generates four off-springs ($os_1, os_2, os_3, os_4 \in \Re^h$) given as follows [8, 24]:

$$os_1 = \frac{w_a^{(g)} + w_b^{(g)}}{2} \tag{6.25}$$

$$os_2 = max(w_a^{(g)}, w_b^{(g)})u + (1 - u)w_{max} \tag{6.26}$$

$$os_3 = min(w_a^{(g)}, w_b^{(g)})u + (1 - u)w_{min} \tag{6.27}$$

$$os_4 = \frac{u(w_a^{(g)} + w_b^{(g)}) + (1 - u)(w_a^{(g)} + w_b^{(g)})}{2} \tag{6.28}$$

The individual maximum and minimum of $w_a^{(g)}$ and $w_b^{(g)}$ are denoted by the vectors $max(w_a^{(g)}, w_b^{(g)})$ and $min(w_a^{(g)}, w_b^{(g)})$ respectively and $u \in [0, 1]$ (here value of u is taken as 0.9) is the crossover weight (closer the value to 1, greater the contribution from the parents). Vectors with max and min possible gene values are represented by w_{max} and $w_{min} \in \Re^h$. The offspring having the smallest fitness value is chosen after the off-springs are produced by the crossover process. The resultant output is represented by the vector which replaces the individual with the greatest fitness in the population (w_{worse}^g replaced with worse $\in [1, M]$).

6.4.2.2   Mutation Process

After the completion of crossover process with mutation probability $mu_p = 0.1$, three new mutated off-springs $mo_1, mo_2, mo_3 \in \Re^h$ are generated by using os_{best} as follows [8, 24]:

$$mo_i = os_{best} + q_i v_i, i = 1, 2, 3 \tag{6.29}$$

the inequalities $w_{min} \leq os_{best} + q_i \leq w_{max}, i = 1, 2, 3$ are satisfied by the vector q_i. Vector v_i falls in the range [0, 1] and it satisfies the condition: in vector v_1, we have a non-zero entry and in v_2, we have a randomly chosen non-zero binary vector and a constant vector 1(containing the value 1 only) is represented by v_3.

A scheme generating a random number between [0, 1] is used for incorporating the generated off-springs in the population where if the random number generated is less than mu_p then $w_{worse}^{(g)}$ the is replaced by the offspring with smallest fitness value,

Otherwise, for $i = 1, 2, 3$ perform the following steps: replace $w_{worse}^{(g)}$ with mo_i if the fitness of mo_i is smaller than $w_{worse}^{(g)}$.

Further, it is to be noted that for the replacement of $w_{worse}^{(g)}$ with any other individual generated, index (worse) must also be updated with the index having larger fitness value from the new population. The stopping criteria used here are as follows: Max generation number $CMGA_{gen} = 8000$, reduction in the training error process (TEP) of the FF, $TEP \leq 10^{-6}$. Algorithm 6.2 shows the steps for DEP(CMGA) algorithm.

Algorithm 2: Steps for DEP(CMGA) Algorithm

---

begin
   DEP(CMGA);
   Data: Population initialized [24] chaotically according to PICO Algorithm 6.1
   Result: Predicted Value
   initialization of CMGA parameters;
   initialization of stopping criteria;
   g=0;
   while *Termination criteria is not satisfied* **do**
      g=g+1;
      for *j=1 to M* **do**
                               ▷ Where M is the population size;
         Initializate DEP parameters taking values from $w_j^{(g)}$;
         Calculate the value of y and the IE for all input patterns.;
         Assess the individual fitness $ff(w_j^{(g)})$ values using Eq. 6.22;
      end
      Select the parents ($w_{a1}^{(g)}$ and $w_{a2}^{(g)}$) from the population;
                             ▷ Where a, b are indices within [1,U];
      for *j=1 to 4* **do**
         Initialization of DEP parameters taking values from os_j;
         Calculate the value of y and the IE for all input patterns;
         Assess the individual fitness $ff(os_j) using Eq. 6.22$;
      end
      The best evaluated offspring is denoted by os_{best};
      for *j=1 to 3* **do**
         Initialization of DEP parameters taking values from mo_j;
         Calculate the value of y and the IE for all input patterns;
         Assess the individual fitness $ff(mo_j)$ using Eq. 6.22;
      end
      os_{best} replaces $w_{worse}^{(g)}$;
      if $Ran_{num} \leq mu_p$ **then**
         Individual among mo_1, mo_2 and mo_3 with smallest fitness value replaces $w_{worse}^{(g)}$;
      else
         for *j=1 to 3* **do**
             if $ff(mo_j) \leq ff(w_{worse}^{(g)})$ **then**
                then mo_j replaces $w_{worse}^{(g)}$
            end
         end
      end
      end
   end
end

6.5 Evaluation Criteria

Many performance criteria are available in literature for prediction evaluation out of which mostly one of the measure is employed known as mean squared error (MSE), which can provide directions to the prediction model [21]. This measure can't be considered as a good performance metric alone for comparing different prediction models.

The model deviation is given by MMRE which is the first metric [8, 21]:

$$MMRE = \frac{1}{N} \sum_{j=1}^{N} \frac{\left|target_j - predicted_j\right|}{target_j} \qquad (6.30)$$

where, the number of input patterns is represented by N, desired output by $target_j$, predicted output by $predicted_j$.

The second metric PRED(s) represents the percentage of prediction which falls within the range of actual values represented by [8, 21]:

$$PRED(s) = \frac{100}{N} \sum_{j=1}^{N} S_j \qquad (6.31)$$

where,

$$S_j = \begin{cases} 1, & if(MMRE_j) < \frac{s}{100} \\ 0, & Otherwise \end{cases} \qquad (6.32)$$

where, s = 25.

A combination of MMRE and PRED called the evaluation function (EF) is used for more accurate results given by [8, 21]:

$$EF = \frac{PRED(25)}{(1 + MMRE)} \qquad (6.33)$$

6.6 Simulations and Results

The proposed model was tested on an agile dataset collected from 21 different software projects from 6 different software industries [11]. For maintaining the large variations of predictions and getting closer values, the normalization of dataset was done following which the dataset was divided into three distinct sets where training set includes 50% of data, validation set includes 25% of data and test set includes 25% of data. The values of a,b were initialized in the interval $[-1, 1]$ and in the range [0, 1] and given as initial input to the PICO algorithm. Comparison of the results obtained here was done with those obtained previously in literature in cascade correlation neural network (CCNN) [12], probabilistic neural network (PNN), group method of data handling polynomial neural network (GMDH-PNN), general regression neural network (GRNN) and bayesian network (BN) [1].

Table 6.1 Prediction results for Zia dataset

Model type	PRED(25)	MMRE	$Eval_f$
PNN Kernel [12]	87.65	1.5776	34.0045
GRNN [12]	85.91	0.3581	63.2574
GMDH-PNN [12]	89.66	0.1563	77.5481
CCNN [12]	94.76	0.1486	82.5004
BN [1]	95.21	0.1369	83.7452
DEP(CMGA)	96.79	0.0565	91.6138

6.6.1 Zia Dataset

Zia dataset [11] was used as a benchmark for evaluating the proposed model which was built with the details collected from 21 software projects from 6 different software industries. The dataset contains 7 independent features: effort (sum of efforts of all individual stories), Vi (units of effort completed in a sprint time by the agile team), D (deceleration which is the rate of negative change of velocity), V (which is the final velocity calculated as Dth power of Vi), sprint size (time set for each iteration and for the schedule), work days and team salary; and one dependent features: actual time. The method of leave one out cross validation (LOOCV) was used for generalization of error [8]. Out of these attributes in the dataset, initially total total count of story points (referred as effort in the dataset) and project velocity (referred as V in the dataset) are considered as input parameters to the CMGA-DEP model in order to estimate their effects over predicted effort output value.

   It can be observed form the above Table 6.1 that BN is the best model found in literature with EF83. But proposed DEP(CMGA) model outperforms the BN model with EF = 91.6138. The proposed DEP(CMGA) model has better performance values in all three performance metrics, PRED(25), MMRE and EF with values 96.79, 0.0565 and 91.6138 respectively. There was an improvement of 2.1422, 61.9784 and 11.0464% with respect to PRED(25), MMRE and EF in the DEP(CMGA) model over the CCNN model which is observed as the second best existing model in the similar working environment. There was an improvement of 1.6594, 58.7289 and 9.3958% with respect to PRED(25), MMRE and EF in the DEP(CMGA) model over the BN model which is observed as the best existing model in the similar working environment. Figure 6.1 shows the graphical comparison of different SDEE methods in terms of EF.

6.7   Conclusion and Future Work

In this paper authors proposed an evolutionary chaotically modified morphological approach for solving SDEE problem. For improving and optimizing the parameters

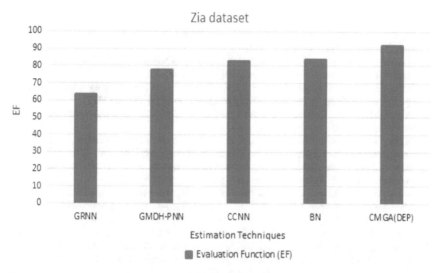

Fig. 6.1   EF obtained for Zia dataset using DEP(CMGA)

of DEP perceptron, CMGA algorithm has been used in this model which results in a higher precision for solving the SDEE problem.

The model was tested on Zia dataset and two metrics were used for evaluating the performance, MMRE and PRED. An evaluation function was used along with the two metrics for creating a global indicator of the performance of the model.

It can be observed that the propose model has better performance with respect to other models and has simpler non-linear components along with fast convergence (better population initialization by PICO algorithm). As per the global indicator EF, DEP(CMGA) showed an improvement of EF = 11.0464% and EF= 9.3958% is observed over CCNN model which is second best model and BN which is best model available for SDEE respectively.

As a future work, extension to this process can be done by making use of other machine learning (ML) techniques including rigorous testing on other agile datasets.

References

1. Dragicevic, S., Turic, S.C.M.: Bayesian network model for task effort estimation in agile software development. J. Syst. Softw. (Elsevier) **127**(1), 109–119 (2017)
2. Bilgaiyan, S., Mishra, S. and Das, M.: A review of software cost estimation in agile software development using soft computing techniques. In: 2nd International Conference on Computational Intelligence and Networks (CINE), pp. 112–117. IEEE (2016)
3. Strode, D.E.: A dependency taxonomy for agile software development projects. Inf. Syst. Front. (Springer) **18**(1), 23–46 (2016)

4. Dingsoyr, T., Moe, N.B., Fagri, T.E., et al.: Exploring software development at the very large-scale: a revelatory case study and research agenda for agile method adaptation. Empir. Softw. Eng. (Springer) **23**(1), 490–520 (2018)
5. Alahyari, H., Svensson, R.B., Gorschek, T.: A study of value in agile software development organizations. J. Syst. Softw. (Elsevier) **125**(1), 271–288 (2017)
6. Hoda, R., Salleh, N., Grundy, J., et al.: Systematic literature reviews in agile software development: a tertiary study. Inf. Softw. Technol. (Elsevier) **85**(1), 60–70 (2017)
7. Dominguez, J.: The Curious case of the chaos report (2009). https://www.projectsmart.co.uk/the-curious-case-of-the-chaos-report-2009.php
8. Araujo, R.A., Oliveira, A.L.I., Soares, S. et. al.: An evolutionary morphological approach for software development cost estimation. Neural Netw. (Elsevier) **32**(1), 285–291 (2012)
9. Araujo, R.A.: A class of hybrid morphological perceptrons with application in time series forecasting. Knowl.-Based Syst. (Elsevier) **24**(4), 513–529 (2011)
10. Banon, G.J.F., Barrera, J.: Decomposition of mappings between complete lattices by mathematical morphology part I. General lattices. Signal Process. (Elsevier) **30**(3), 299–327 (1993)
11. Zia, Z.K., Tipu, S.K., Zia, S.K.: An effort estimation model for agile software development. Adv. Comput. Sci. Its Appl. (World Sciences) **2**(1), 1–6 (2012)
12. Panda, A., Satapathy, S.M., Rath, S.K.: Empirical validation of neural network models for agile software effort estimation based on story points. In: 3rd International Conference on Recent Trends in Computing, pp. 772–781. Elsevier (2015)
13. Reeves, C.: Genetic algorithms. Handbook of Metaheuristics (Springer) **57**(1), 55–82 (2011)
14. Sharma, A., Bawa, R.K.: A roadmap for agility estimation and method selection for secure agile development using AHP and ANN. Data Eng. Intell. Comput. (Springer) **542**, 237–245 (2017)
15. Araujo, R.A., Soares, S., Oliveira, A.L.I.: Hybrid morphological methodology for software development cost estimation. Expert. Syst. Appl. (Elsevier) **39**(1), 6129–6139 (2012)
16. Braga, P.L., Oliveira, A.L.I., Meira, S.R.L.: Software effort estimation using machine learning techniques with robust confidence intervals. In: International Conference on Tools with Artificial Intelligence (ICTAI), vol. 8, pp. 1595–1600. IEEE (2007)
17. Oliveira, A.L.I., Braga, P.L. et. al.: GA-based method for feature selection and parameters optimization for machine learning regression applied to software effort estimation. Inf. Softw. Technol. (Elsevier) **52**(1), 6129–6139 (2010)
18. Araujo, R.A, Oliveira, A.L.I., Soares S.C.B.: A morphological-rank-linear approach for software development cost estimation. In: 21st IEEE International Conference on Tools with Artificial Intelligence, pp. 630–636 (2009)
19. Araujo, R.A., Oliveira, A.L.I., Soares, S.: Gradient based morphological approach for software development cost estimation. In: Proceedings of the Symposium on Applied Computing, pp. 588–594. IEEE (2011)
20. Araujo, R.A., Oliveira, A.L.I., Soares, S.: Hybrid intelligent design of morphological-rank-linear perceptrons for software development cost estimation. In: 22nd International Conference on Tools with Artificial Intelligence, pp. 160–167. IEEE (2010)
21. Araujo, R.A., Oliveira, L.I., Soares, S.: A shift-invariant morphological system for software development cost estimation. Expert. Syst. Appl. (Elsevier) **38**(4), 4162–4168 (2011)
22. Choudharia, J., Suman, U.: Story points based effort estimation model for software maintenance. Procedia Technol. (Elsevier) **4**, 761–765 (2012)
23. Estevao, P.S., Esmi, L.: Morphological perceptrons with competitive learning: lattice-theoretical framework and constructive learning algorithm. Inf. Sci. (Elsevier) **181**(10), 1929–1950 (2011)
24. Leung, F.H.F., Lam, H.K., Ling, S.H.: Tuning of the structure and parameters of a neural network using an improved genetic algorithm. IEEE Trans. Neural Netw. **14**(1), 79–88 (2003)
25. Gao, W., Liu, S., Huang, L.: Particle swarm optimization with chaotic opposition-based population initialization and stochastic search technique. Commun. Nonlinear Sci. Numer. Simul. **17**(11), 4316–4327 (2012)

Chapter 7
PerbDroid: Effective Malware Detection Model Developed Using Machine Learning Classification Techniques

Arvind Mahindru and A. L. Sangal

Abstract This chapter introduces PerbDroid—a framework to detect malware from Android smartphones. To address the issues of malware detection through a broad set of apps, researchers have recently started to identify the features which helps to detect malware from apps. The proposed framework is based on features selection techniques which help us to develop a useful model for malware detection. We collected a data set of 2,00,000 Android apps from distinct sources and extracted permissions and API calls from them (consider as features in this study). Further, features are selected by using six different feature ranking approaches (i.e., Gain Ratio, OneR feature evaluation, Chi-squared test, Information gain feature evaluation, Principal Component Analysis (PCA) and Logistic regression analysis) to develop the model for malware detection. We evaluated several machine learning algorithms and feature selection methods in identifying the combination that gives the foremost performance to detect malware from real-world apps. Empirical outcomes illustrate that the proposed framework is useful to detect malware from smartphones mainly and in particularly from Android.

7.1 Introduction

In the digital world, Smartphones play the role of computers, and it can be used anywhere and anytime by anyone. While making communication easy, Smartphones can provide all types of facilities and apps, without which, many individuals will be lost and incapable to survive for their daily traditions. Such facilities and apps consist of Map, social media, e-banking, e-mail, online games, and many more conveniences. While using these conveniences in smartphones, the user have to permit several

A. Mahindru (✉) · A. L. Sangal
Department of Computer Science and Engineering, Dr. B.R. Ambedkar National Institute of Technology, Jalandhar 144001, India
e-mail: er.arvindmahindru@gmail.com

A. L. Sangal
e-mail: sangalal@nitj.ac.in

© Springer Nature Switzerland AG 2020
J. Singh et al. (eds.), *A Journey Towards Bio-inspired Techniques in Software Engineering*,
Intelligent Systems Reference Library 185,
https://doi.org/10.1007/978-3-030-40928-9_7

permissions (i.e., features) to these apps by which Cybercriminals take control over their Smartphones.

In the first half of 2019, the share of mobile devices with the Android operating system is 84.4% throughout the world.[1] Due to the rapid growth in smartphone sales, it attracts the attention of cybercriminals to develop malware-infected apps and steal private user information on daily basis. According to the report issued by Kaspersky Lab at the end of 2018[2] they detect approximately 5,321,142 Android malware samples belonging to different families. Studies [1, 2] show that, this malware problem will increase day by day which is a threat to user's privacy. To overcome this issues, the smartphone user attempts to defend the smartphones by upgrading the apps and anti-virus software. So, that smartphone user must give consistent attention to the permissions demand by the apps. Permissions that are requests for contacts sharing, location sharing, SMS sharing, and sensitive information should be approved with cautions and care.

Previous studies revealed that there are several detection techniques available to detect Android Malware [3, 4]. For Android Malware detection, static and dynamic analysis techniques were developed in the past [5]. These approaches have earned popularity due to its effectiveness in mitigate the effect of malware. In static analysis technique, malware is detected through its App Program Interface (API) [6], Permissions [7], and its Code structure [8]. However, dynamic analysis can identify malware, based on its normal or abnormal operations seen in an isolated environment [5]. If the app, demand normal permissions,[3] then it belongs to a benign group, and if it demands dangerous permissions,[4] then it belongs to the malware group. Both dynamic and static analysis approaches are essential to discover malware from Android apps [9]. However, because of the open source of Android operating system, smartphones have become an easily target for cybercrooks who intrusion the smart devices by uploading their malware app as a benign app in the Google play store. This concerns has become a serious issue for mobile users because it can steal user information. Whenever the user installs a new app, they are required to permit certain permissions for its proper functioning. However, the user who are going to install apps do not know that the permission which is required by the apps are dangerous or normal. To protect smartphones from cyber criminals smartphone users would need good knowledge of permissions [10]. Further, malware apps, when used, also require the same permissions as the benign apps require for their functionalities. Therefore, users are using malware apps unknowingly. To solve this problem, a practical approach is needed which can achieve high detection accuracy rate by consuming less memory and time.

In this chapter, we propose a framework which works on the principle of feature selection approaches to select the foremost features for Android malware detection

---

[1] http://gs.statcounter.com/android-version-market-share/mobile-tablet/worldwide.

[2] https://securelist.com/mobile-malware-evolution-2018/89689/.

[3] https://developer.android.com/guide/topics/permissions/normal-permissions.html.

[4] https://developer.android.com/guide/topics/permissions/requesting.html.

Fig. 7.1 Flow chart of the propose work

which helps us to achieve higher accuracy, less power, and memory consumption. The novel and unique contributions of this chapter are presented as follows:

- To develop a malware detection model involving a set of features suitable for all categories of Android apps.
- A performance detection framework to examine the performance of our proposed malware detection model.

The phases followed by us in building an effective Android malware detection model is demonstrated in Fig. 7.1. Initially, at the very first phase of malware detection we collect Android application packages (.apk) from distinct sources. Next, the class of .apk file is identified by using different antivirus scanners. Further, a different set of features are extracted by using distinct tools available in the literature. Next, the right set of features are chosen using a feature selection approach. In the next phase, these selected features are recognized as input to build a model by utilizing distinct classification approaches. Finally, the best suitable model is used to validate the proposed detection framework.

The remaining sections of the chapter are framed as follows: Sect. 7.2, present the literature survey for Android malware detection and the research questions. In Sect. 7.3, we present the description of the data set. In Sect. 7.4, we explain the feature selection approaches. Section 7.5, present the machine learning classifiers utilized in our study. In Sect. 7.6, we explain the proposed framework to check that proposed model is able to detect malware or not. Section 7.7, presents the evaluation parameters used in our study. In Sect. 7.8, we propose our Android malware detection framework PerbDroid. Section 7.9, represents the results that how much our proposed framework is capable to detect malware from real-world apps, experimental findings, conclusion and future work of this chapter.

7.2 Related Work

This part of the chapter contains the previous work done to detect malware from real-world apps. Table 7.1 present the existing literature in the area of malware detection. The relationship between features and the malware detection acquired from the available research is indicated in this table. First column of this points towards the title of the approach or author which exist in the literature for Android malware detection. Second column points towards the data set used by the researchers for malware detection. Third column of the table depicts the approaches used by the researchers

Table 7.1 Existing literature having data set, method used and features used to detect malware

Approach/author	Data set used	Classification algorithm used	Feature utilized
DroidMat 2012 [11]	1,738	Naïve Bayes and KNN	Permissions and API calls
AndroSimilar 2013 [12]	8,584	Fuzzy hashing approaches	Entropy features for byte blocks
Wu et al. 2016 [13]	2,210	KNN	API calls
Qiao et al. 2016 [14]	6,260	Support vector machines, Random forest, and artificial neural networks	Permission and API calls
Wang et al. 2017 [15]	36,964	SVM	HTTP
DroidDet 2018 [16]	2,130	Ensemble rotation forest	Permissions, API calls and permission rate
Vinayakumar et al. 2018 [17]	1,738	LSTM-RNN	Permissions
Shen et al. 2018 [18]	8,598	Gram based classification	API calls

for malware detection. The fourth column tells the set of features recognized while building a model to detect either the app is benign or malware. Table 7.1 indicates that the efficiency of the malware detection model relies on the set of features which have been recognized as input to build a model. In the present chapter, a research on the impact of distinct sub-sets of features (a set consisting of all the extracted features and six reduced sets of features determined by utilizing feature selection approaches) on the property of attributes.

In 2013, Faruki et al. [12] suggested AndroSimilar, a persistent statistical signature approach for detecting the unknown versions of already present malware that are normally developed by utilizing code obfuscation and repackaging approaches. They tested the proposed framework AndroSimilar by using 1260 apps. Haung et al. [19] applied four machine learning algorithms on the data set of 480 malware apps and 124,769 benign and get 81% of malware apps can be detected on the basis of permission-based detection approach. Sanz et al. [20] proposed PUMA for detection of malware apps by analyzing the demanded permission for the app. They implemented distinct classifier algorithms on the data set of 249 malware and 357 benign apps and got high detection rates and false positive rate too. Shin et al. [21] employed machine-based technique and properly analyzed the Android permission-based security model. Wei et al. [22] suggested a security distance model to mitigate the Android malware. Security distance model is established upon the notion that not a single permission is sufficient for an Android app to threat the assurance of Android based devices. It measures the dangerous level of apps based on permission they required. Authors, divide the permissions in four categories, which helps the user to find out which apps are more threatening to malware. Enck et al. [23] build a tool KIRIN that gives lightweight certification at the time of installation. It deter-

Table 7.2 Summary of literature to detect Android malware using different feature selection techniques

Approach/author	Feature selection technique used
Andromaly [25]	Fisher score, Chi-square and information gain
Mas'ud et al. [27]	Information gain and Chi-square
MKLDroid [28]	Chi-squared
Allix et al. [2]	Information gain
Azmoodeh et al. [29]	Information gain

mines the security rules and only matches the permissions demand by the apps at the time of installation and certify the app to malware if it does not pass the defined rules. Wu et al. [11] proposed DroidMat that work on the principle of extraction the data from manifest files such as permissions, message passing and API call tracing to study the behavior of the app by using the KNN algorithm.

Iker et al. [24] proposed a dynamical analysis tool named as Crowdroid to detect the behavior of Android apps. K-mean clustering algorithms were applied on Android apps to identify that app is malware or benign. Shabtai et al. [25] build a behavior-based Android malware detection approach named as Andromaly. It constantly monitors the distinct patterns and features that show the device states and then apply machine learning algorithms to identify which app is malicious or benign. Huang et al. [26] proposed a type-based taint analysis for Android named as DFlow, a context-sensitive information flow type approach. They propose novel approaches for error reporting which is dependent on the CFL-reach ability, as well as novel approaches for handling of Android-specific features, which include libraries, multiple entry points, and call-backs and inter-component communication.

Our observations suggest that most authors commonly use logistic regression, decision tree analysis, neural network and Naïve Bayes classifier. In this chapter, we consider ten different classifiers to design a model which help to detect either the app is benign or malware.

Earlier, researchers also presented feature selection techniques for detecting the effectiveness of malware detection. Table 7.2 highlights on the research conducted by different authors which is based on feature selection and is used to detect malware from Android apps. In this work, we present the detection evaluation framework to measure the effectiveness of build malware detection models by considering thirty different sets of the features as valuable input.

7.2.1 Research Questions

An empirical study is performed to discover out the efficacy of malware detection approaches by utilizing the proposed detection framework. Our study also emphasizes on determining the best possible set of features to detect either the app is malware or benign. Following are the research questions observed from the previous research gaps:

RQ1: Which malware detection model is most appropriate among the distinct techniques?

This question assist to examine the performance of the distinct classifiers which are used to detect malware from Android apps. In this chapter, we implement ten distinct kinds of classifiers for developing a model by recognizing a set of features as input and compare the outcome of the model which is better to detect either the app is benign or malware.

RQ2: Do the feature selection approaches pay any impact on the functioning of the classification approaches?

It is seen that few of the feature selection approaches worked so accurately with a particular classification approach. Therefore, in the current chapter, we evaluate six distinct ranking approaches by utilizing distinct classification approaches to measure their performance. This question mainly addresses the variance of performance of the classification model over distinct classification models.

RQ3: Which feature ranking method perform better for the task of detecting malware from Android apps?

This question helps us to select the best features by using the ranking methods. Further, best-ranked features are used to develop a model to detect either the app is benign or malware.

RQ4: Does a selected set of features perform better than considering all set of features for the task to detect whether the app is benign or malware?

In this research question, our objective is to review the set of features and study the relationship with benign or malware apps. In this chapter, six distinct kinds of feature ranking approaches have been recognized for identifying a sub-set of features which are better to detect whether the app is benign or malware.

7.3 Data Set Description

From Table 7.1, we have seen that most of the researchers considered a limited range of Android apps to study the relationships among malware and set of features. Consequently, we cannot draw any conclusions applicable to all categories of Android apps and systems in general (i.e., Android in our case). So in this study, we consider thirty different categories of Android apps are recognized to generalize and strengthen our outcomes. We collect the experimental data set from promise repositories. We collected 2,00,000 of .apk files, from Google's play store,[5] appchina,[6] hiapk,[7] Android,[8]

---

[5]https://play.google.com/store?hl=en.

[6]http://www.appchina.com/.

[7]http://www.hiapk.com/.

[8]http://andrdoid.d.cn/.

mumayi,[9] gfan,[10] pandaapp[11] and slideme.[12] Among these 1,95,000 benign .apk files, 1,75,000 are distinct. These apps are collected after removing viruses, reported by Microsoft Windows Defender[13] and Virus Total.[14] Virus Total[15] helps to detect malware by antivirus engines and includes over 60 antivirus software. A number of 35,000 malware samples, from three different datasets [30, 31] are collected. In [30], Kadir et al. introduced an Android sample set of 1929 botnets, consisting of 14 different botnet families. Android Malware Genome project [31] contains a collection of 1200 malware samples that cover the majority of existing Android malware families. We collected about 17,871 samples from AndroMalShare[16] along with their package names. After removing duplicate packages from the collected dataset, we have 25,000 different malware samples left in our study. Both malware and benign apps have been collected from the sources mentioned above until March 2019. Table 7.3 gives the category of the Android app with the number of samples used in our study.

Next phase, is to extract the features from collected .apk files to develop a model for malware detection. To extract features from the collected .apk files, we use Android studio as an emulator and extract features (i.e., permissions and API calls) by self-written code in Java language from the collected Android apps. Further, the extracted features are divided into thirty different sets. Table 7.4 show us the formulation of different sets that contains the information about the features (i.e., permissions, API calls, number of user download the app, and the rating of the app).

7.4 Feature Selection Approaches

It is necessary to choose appropriate set of features for data preprocessing task in machine learning. On the basis of Table 7.2, it is seen that in the previous studies, researchers applied distinct feature ranking methods to choose the best features to detect malware from Android apps. In this chapter, six distinct kinds of feature ranking approaches are implemented on thirty distinct categories of Android app to discover the best set of features which support us to detect malware with better accuracy and also minimize the value of misclassification errors. The subsequent subsections emphasizes on distinct feature ranking approaches to discover a small set of features from total possible features that jointly have excellent detection capability.

---

[9]http://www.mumayi.com/.

[10]http://www.gfan.com/.

[11]http://android.pandaapp.com/.

[12]http://slideme.org/.

[13]https://www.microsoft.com/en-in/windows/comprehensive-security.

[14]https://www.virustotal.com/.

[15]See footnote 14.

[16]http://sanddroid.xjtu.edu.cn:8080.

Table 7.3 Categories of .apk files belong to their respective families (*.apk*)

ID	Category	N	T	Ba	W	Bo	S
D1	Arcade and Action (AA)	6291	440	100	204	130	600
D2	Books and Reference (BR)	5235	200	250	56	150	150
D3	Brain and Puzzle (BP)	4928	820	54	28	50	50
D4	Business (BU)	8308	152	150	150	22	22
D5	Cards and Casino (CC)	2886	76	65	81	100	44
D6	Casual (CA)	2010	321	69	46	150	140
D7	Comics (CO)	7667	65	95	35	3	0
D8	Communication (COM)	8414	250	50	500	3	3
D9	Education (ED)	8744	560	68	50	50	68
D10	Entertainment (EN)	4222	500	500	500	100	42
D11	Finance (FI)	3999	50	200	99	65	92
D12	Health and Fitness (HF)	8551	98	65	45	140	140
D13	Libraries and Demo (LD)	5655	70	100	100	6	500
D14	Lifestyle (LS)	7650	155	200	100	193	192

(continued)

Table 7.3 (continued)

ID	Category	N	T	Ba	W	Bo	S
D15	Media and Video (MV)	8019	100	123	162	450	71
D16	Medical (ME)	6000	12	13	12	24	25
D17	Music and Audio (MA)	8621	65	100	65	165	165
D18	News and Magazines (NM)	8164	100	100	100	100	32
D19	Personalization (PE)	4334	500	42	500	200	22
D20	Photography (PH)	9133	100	120	50	96	500
D21	Productivity (PR)	9850	100	516	250	250	62
D22	Racing (RA)	7766	50	100	210	100	180
D23	Shopping (SH)	2673	100	100	120	150	50
D24	Social (SO)	6159	100	50	210	150	150
D25	Sports (SP)	2669	100	240	100	450	112
D26	Sports Games (SG)	3889	100	145	145	650	198
D27	Tools (TO)	3346	120	500	550	475	563
D28	Transportation (TR)	3796	2	500	100	100	20
D29	Travel and Local (TL)	3180	500	220	150	48	100
D30	Weather (WR)	2841	120	23	700	50	25

a"N" stands for *Normal*, "T" stands for *Trojan*, "Ba" stands for *Backdoor*, W stands for "Worm", "BO" stands for *Botnet*, and "S" stands for "*Spyware*"

Table 7.4 Formulation of sets (having permissions, API calls, number of user download the app and rating of the apps)

Set no.	Description	Set no.	Description
S1	SYNCHRONIZATION _DATA	S2	CONTACT_INFORMATION
S3	PHONE_STATE and PHONE_CONNECTION	S4	AUDIO and VIDEO
S5	SYSTEM_SETTINGS	S6	BROWSER_INFORMATION
S7	BUNDLE	S8	LOG_FILE
S9	LOCATION_INFORMATION	S10	WIDGET
S11	CALENDAR_INFORMATION	S12	ACCOUNT_SETTINGS
S13	DATABASE_INFORMATION	S14	IMAGE
S15	UNIQUE_IDENTIFIER	S16	FILE_INFORMATION
S17	SMS_MMS	S18	READ
S19	ACCESS_ACTION	S20	READ_AND_WRITE
S21	YOUR_ACCOUNTS	S22	STORAGE_FILE
S23	SERVICES_THAT_COST_ YOU_MONEY	S24	PHONE_CALLS
S25	SYSTEM_TOOLS	S26	NETWORK_INFORMATION and BLUETOOTH_INFORMATION
S27	HARDWARE_CONTROLS	S28	DEFAULT GROUP
S29	API CALLS	S30	RATING and USER DOWNLOADS THE APP

7.4.1 Chi-Squared Test

Chi-squared test is employed to examine the self-determination among two events [32], here, grading of features are predicted on the significance of the Chi-squared statistic in relation to the class. Higher value of Chi- squared imply the denial of the null hypothesis and consequently these features can be analyzed as good relevance.

7.4.2 OneR Feature Evaluation

In oneR feature selection approach, is used for grading of features [33]. OneR classifier utilizes classification rates to grade the features. It studied all numerically valuable features as constant ones and divide the set of values into a few dissociate intervals made by straightforward method. Features with better classification rates are studied to be of addition importance.

7.4.3 Gain Ratio Feature Evaluation

In gain ratio feature selection approach, grading of features are predicted on the measures of the gain ratio in relation to the class [33]. The gain ratio of feature 'Z' is determined as:

$$\text{Gain Ratio} = \frac{Gain(Z)}{SplitInfo_Z(X)},$$ (7.1)

where $Gain(Z) = I(X) - E(Z)$ and X depicts the set including m numbers of instances with n different classes. The forthcoming information necessary to categorize a given sample is calculated by utilizing following equation:

$$I(X) = -\sum_{i=1}^{m} P_i \log_2(p_i).$$ (7.2)

Here in this equation P_i is the chance that a random sample can be a member of class C_i and is measured by n_i/n.

Suppose n_{ij} presents the number of instances of class C_i in sub-set N_j. The expected information is relying on the partition of sub-sets by F, and is presented by

$$E(Z) = -\sum_{i=1}^{M} I(X) \frac{n_{1i} + n_{2i} + \cdots + n_{mi}}{n}.$$ (7.3)

The value of $SplitInfo_F(Z)$ is measured by using following equation:

$$SplitInfo_F(N) = -\sum_{i=1}^{t} \frac{|N_i|}{N} \log_2 SplitInfo_F(N)$$ (7.4)

The value of $SplitInfo_F(X)$ show us the information achieved by dividing the training data set X into t partitions equivalent to t results of a test on the attribute Z.

7.4.4 Information Gain Feature Evaluation

In information gain feature selection, significance of features are planted on the basis of the information gain in relation to the class [33].

7.4.5 Logistic Regression Analysis

Logistic regression analysis is a statistical review approach [34]. For this study, Univariate Logistic Regression (ULR) analysis has been recognized to verify the degree of significance for each set of features. In the current work, we considered two benchmarks of LR model which recognized to discover the level of importance of every set of feature and also utilized to grade the features set. These benchmarks are:

1. *Importance of regression coefficient:* The coefficient measure of features indicates the degree of correlation of every set of features with malwares.

2. *P-value:* P-value i.e., implication level indicates the significance of correlation.

7.4.6   Principal Component Analysis (PCA)

Attribute reduction is achieved by utilizing PCA transform from highest dimension data space into lowest dimension data space. The lowest dimension data have the extremely important features [35]. Considering correlation among several features are highest, so PCA is applied to relocate these features into novel features that are extremely correlated. The recent features are named as principal component domain features. A few principal components (PCs) are recognized enough to depict most of the important standards in the data.

7.5   Machine Learning Classifiers

The presented framework utilize a Machine Learning classification technique for appreciating a Malware detection system. In the present study, the malware detection framework constantly monitors several features and incidents achieved from the system and then implement ten different machine learning classifiers to classify collected observations whether the app is malware or benign. On the basis of earlier studies and after weighing the resource consumption issues, we determine to examine the subsequent classifiers: (i.e., SVM [36], Naïve Bayes [25], Random forest [20], Multiple layer perceptron [27], Logistic regression [25], Bayesian network [25], Adaboost, Decision tree [25], KNN [6], and Deep Neural Network [37]).

Evaluation of machine learning classifier is typically split into two subsequent phases, i.e., testing and training. In the testing phase, we trained the dataset with selected features which are obtained by different feature selection techniques. These features are extracted from both benign and malware apps. In the testing time, the outcomes of the classifier is measured by using selected two performance parameters F-measure and Accuracy.

7.6   Proposed Detection Framework

To measure the performance of our proposed malware detection model. We compare the result of our proposed model with two different techniques.

Table 7.5 Confusion matrix to classify a Android app is benign or malware (*.apk*)

	Benign	Malware
Benign	Benign->Benign	Benign->Malware
Malware	Benign->Malware	Malware->Malware

a. Comparison with AV scanners: To compare the outcome of our proposed model, we select five different anti-virus scanners and compare their performance with our proposed framework.

b. Comparison with previously used classifiers: To check the feasibility of our proposed model, we compare the parameters like F-measure and Accuracy with other models build in the literature.

7.7   Parameters Used for Evaluation

The subsequent sub-sections yield the fundamental descriptions of the performance parameters utilized for malware detection. All these parameters are calculated by utilizing Confusion matrix. Confusion matrix includes actual and detected classification information done by detection approach. Table 7.5 gives the confusion matrix for the malware detection model. For our work, two performance parameters F-measure and Accuracy are utilized for evaluating the performance of malware detection methods.

$$Accuracy = \frac{N_{Benign \rightarrow Benign} + N_{Malware \rightarrow Malware}}{N_{classes}}.$$

$$F-measure = \frac{2 * Precision * Recall}{Precision + Recall}$$
$$= \frac{2 * N_{Malware \rightarrow Malware}}{2 * N_{Malware \rightarrow Malware} + N_{Benign \rightarrow Malware} + N_{Malware \rightarrow Benign}}.$$

7.8   Setup to Perform Experiment

In this part of the paper, the experimental set-up to discover the efficacy of malware detection model using the proposed detection framework is presented. SVM [36], Naïve Bayes [25], Random forest [20], Multiple layer perceptron [27], Logistic regression [25], Bayesian network [25], Adaboost, Decision tree [25], KNN [6], and Deep Neural Network [37] are utilized to build a model that detect either the app is benign or malware. These approaches are implemented on thirty different

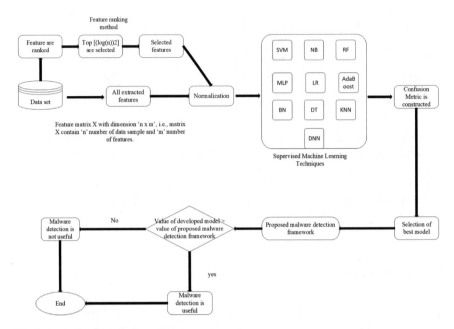

Fig. 7.2 Framework of proposed work

categories of Android apps, as shown in Table 7.3. All these categories have a different percentage of benign or malware apps. Figure 7.2 shows the proposed framework.

The subsequent steps are pursued for selecting a set of features to built the malware detection model which is able to detect either the app is benign or malware. Feature selection approaches (i.e., feature ranking) are applied to thirty different categories of Android apps. Consequently, a total of 2100 ((6 feature ranking approach + 1 recognizing all features) × 30 different Android apps data sets × 10 detection approaches) different detection models have been build in this work.

1. In this chapter, six feature ranking approaches are applied to thirty different categories of Android apps to choose the appropriate set of features for malware detection.
2. The sub-sets of features achieved from the above mentioned step are used as an input to train ten distinct machine learning algorithms while building a model. To achieve better outcomes, 20-fold cross-validation is utilized for comparing the models, i.e., data sets are partitioned into 20 parts. The performance of all build malware detection models are matched by utilizing two distinct performance parameters F-measure and Accuracy.
3. The models build from above mentioned two steps are passed to the proposed detection framework to validate either the build malware model is useful for this study or not.

Table 7.6 Naming
conventions for distinct
approaches (*.apk*)

Abbreviation	Corresponding name
DS	Data set
FR1	Chi-squared test
FR2	Gain ratio feature evaluation
FR3	Filtered subset evaluation
FR4	Information gain feature evaluation
FR5	Logistic regression analysis
FR6	Principal Component Analysis (PCA)
AF	All extracted features

7.9   Outcomes

In the present part of the chapter, the relationship among different set of features
and malware detection at the Android level is presented. F-measure and Accuracy
are recognized as performance assessment parameters to compare the performance
of malware detection model build by utilizing ten distinct classifiers approaches. To
depict the outcomes, we utilize the following abbreviations as revealed in Table 7.6.
Table 7.6 gives the corresponding of these abbreviations to their authentic names.

7.9.1   Feature Ranking Approaches

In the present work, six feature ranking approaches: information gain, gain ratio
feature evaluation, Chi-squared test, information gain feature evaluation, oneR feature evaluation, logistic regression analysis and principal component analysis are
implemented on a distinct set of features. Each approach utilize distinct performance
parameters to grade the feature. Further top $\lceil \log_2 n \rceil$ set of features out of "n" number
of features have been measured to build a model for detecting malware detection. For
initial four feature ranking approaches (Gain ratio feature evaluation, Information
gain, Chi-squared test and OneR feature evaluation), top $\lceil \log_2 n \rceil$ are chosen as a
sub-set of features, where n is the number of the set of features in the original data
set (for this work $n = 20$). However, in univariate logistic regression (ULR) study,
just those features are chosen which have a positive value of regression co-efficient,
p-value measure is less than 0.05, and in matter of PCA, only those features are
chosen which have Eigenvalue more than 1. The chosen set of features after feature
ranking approaches are demonstrated in Fig. 7.3.

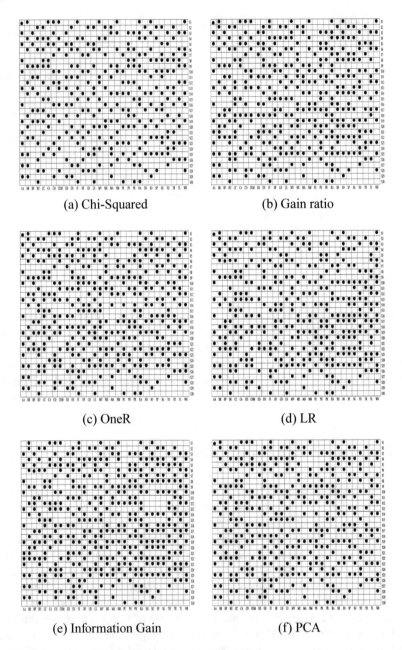

Fig. 7.3 Selected set of features using feature ranking methods

7.9.2 Machine Learning Classifier

In this chapter, ten different classifiers have been recognized to build the requisite model. Seven sub-sets of features (6 selecting from feature selection approaches + 1 recognizing all set of features) are recognized as input to build a model for detecting malware from Android apps. Hardware utilized to conduct this research is: Core i7 processor into 8GB RAM and storage capacity of 1 TB hard disk. Detection models are build by utilizing the MATLAB environment. The outcome of each detection model is measured in terms of two performance parameters, F-measure and Accuracy.

Tables 7.7, 7.8, 7.9, 7.10, 7.11 and 7.12 demonstrate the achieved performance values for distinct data sets by utilizing ten distinct classifiers. On the basis of Tables 7.7, 7.8, 7.9, 7.10, 7.11 and 7.12, it can be implicit that:

- In case of DNN, malware detection model build by considering the chosen set of features by utilizing FR6, i.e., PCA achieved higher results when matched to different feature selection approaches.

In the present chapter, ten distinct classifier and two performance parameters are recognized to detect either the app is benign or malware. Twelve box-plot diagrams are demonstrated in Fig. 7.4 (one for each combination). Each of them contains ten box-plots: one for all machine learning approach. The model which is having few numbers of outliers and high median value is the foremost model to detect either the app is benign or malware. On the basis of the box-plot diagram, it can be concluded that:

- Amongst all implemented feature ranking approaches, FR6 have fewer outliers along with high median value. On the basis of box-plots demonstrated in Fig. 7.4, FR6 generates the best result, i.e., feature ranking by utilizing PCA calculates the best set of features for detecting malware and benign apps and give better results as matched to others.
- Amongst all classifiers approaches implemented on distinct Android data sets, model build using DNN as classifiers give better results as matched to others.

Remark We name our new proposed model as PerbDroid which consist of feature ranking approach PCA and DNN as an classifier.

7.9.3 Comparison of Results

Pair-wise *t*-test is utilized to identify, out of which classifiers and the feature selection approaches work better or all have carried out equally well.

Table 7.7 Accuracy and F-measure using FR1

ID	Accuracy										F-measure									
	SVM	NB	RF	MLP	LR	BN	AB	DT	KNN	DNN	SVM	NB	RF	MLP	LR	BN	AB	DT	KNN	DNN
D1	83.33	85.0	87.67	87.66	82	87	80	86.6	87.6	**88**	0.81	0.82	0.83	0.83	0.85	0.88	0.87	0.86	0.81	**0.89**
D2	80	82.08	86.27	82.66	88	86	83	86.6	87.6	**89**	0.88	0.82	0.84	0.83	0.82	0.86	0.87	0.86	0.86	**0.89**
D3	82	84.8	84.67	81.06	83	85	82	81.6	85.6	**89.7**	0.81	0.80	0.81	0.80	0.81	0.82	0.81	0.85	0.86	**0.88**
D4	82	81.08	82.27	81.60	86	85	82	85.6	87.6	**89**	0.83	0.80	0.82	0.81	0.80	0.82	0.86	0.86	0.87	**0.88**
D5	76	81	82	82	82	**86**	81	86	81	82	0.81	0.77	0.82	0.81	0.80	**0.83**	0.81	0.82	0.81	0.80
D6	80	82.08	81.27	81.66	82	81	83	82.6	82.6	**84**	0.88	0.82	0.84	0.83	0.82	0.86	0.87	0.86	0.86	**0.88**
D7	82	85	87	82	84	81	80	81	83	**86**	0.81	0.80	0.82	0.81	0.80	0.81	0.83	0.83	0.81	**0.86**
D8	79	72	78	81	80	**89**	82	83	86	86	0.81	0.80	0.82	0.80	0.80	**0.89**	0.83	0.81	0.84	0.85
D9	86	88	88	86	85	**87**	87	82	82	85	0.89	0.81	0.84	0.83	0.82	**0.87**	0.89	0.80	0.82	0.81
D10	78	81	82	81	82	81	86	87	87	**88**	0.82	0.81	0.80	0.82	0.83	0.83	0.86	0.80	0.82	**0.87**
D11	82	83	84	83	87	82	81	84	83	**85**	0.80	0.81	0.82	0.82	0.83	0.82	0.81	0.80	0.81	**0.84**
D12	80	81	82	82	81	81	83	82	83	**85**	0.82	0.80	0.81	0.81	0.81	0.82	0.83	0.82	0.81	**0.88**
D13	78	81	81	82	82	81	82	86	86	**89**	0.78	0.80	0.81	0.80	0.79	0.81	0.82	0.80	0.81	**0.88**
D14	81	80	86	82	85	82	81	86	86	**88**	0.82	0.82	0.81	0.80	0.86	0.83	0.85	0.82	0.82	**0.88**
D15	82	83	83	82	84	**87**	84	81	80	82	0.82	0.84	0.83	0.82	0.83	**0.87**	0.84	0.82	0.83	0.84
D16	71	80	87	86	82	83	81	86	87	**88**	0.82	0.81	0.80	0.80	0.81	0.83	0.80	0.80	0.81	**0.83**
D17	81	81	82	86	89	82	83	86	81	**88**	0.82	0.80	0.83	0.85	0.86	0.87	0.87	0.87	0.81	**0.88**
D18	82	84	81	80	81.8	82	82.9	85	86	**89**	0.83	0.81	0.83	0.86	0.83	0.85	0.86	0.83	0.84	**0.87**
D19	86	87	89	81	82	**88**	84	85	86	86	0.81	0.82	0.83	0.85	0.82	**0.86**	0.82	0.80	0.81	0.88
D20	82	86	86	84	87	81	83	**88**	81	87	0.82	0.86	0.83	0.82	0.85	0.87	0.88	**0.89**	0.86	0.87
D21	79	78	80	82	87	86.7	89	87	89	**89.7**	0.82	0.85	0.84	0.85	0.87	0.88	0.83	0.87	0.81	**0.89**
D22	88	86	87	89.7	89.3	81	82	84	85	**88**	0.82	0.83	0.84	0.85	0.82	0.82	0.85	0.84	0.86	**0.88**
D23	80	81	80.88	83.76	88.78	89.71	82.9	**88.9**	83	84	0.72	0.80	0.81	0.80	0.82	**0.85**	0.84	0.82	0.80	0.81

(continued)

Table 7.7 (continued)

	Accuracy										F-measure									
D24	81	80	80.8	82	81	**87.2**	81.3	80	82.7	86	0.80	0.81	0.82	0.86	0.85	**0.89**	0.82	0.82	0.85	0.85
D25	86	88	86	89	82	82	83	86	81	**89**	0.83	0.81	0.86	0.87	0.83	0.83	0.81	0.80	0.82	**0.88**
D26	84	87.7	86.9	86.6	80.8	82.1	84	87	85	**88**	0.80	0.81	0.82	0.83	0.84	0.81	0.82	0.83	0.84	**0.85**
D27	82	85	82	81	80.1	81.9	83.6	83	86.9	**87.9**	0.80	0.82	0.81	0.86	0.82	0.85	0.85	0.86	0.87	**0.88**
D28	86	88	86	82	80	82	83	84	82	**89**	0.78	0.82	0.85	0.82	0.85	0.83	0.82	0.88	0.85	**0.89**
D29	78	82	85	86	89	81	80	85	84	**88**	0.80	0.81	0.84	0.86	0.82	0.82	0.86	0.87	0.85	**0.88**
D30	76	82	84	87	89	81	82	85	81	**88**	0.78	0.81	0.82	0.85	0.86	0.87	0.85	0.87	0.87	**0.88**

Table 7.8 Accuracy and F-measure using FR2

ID	Accuracy										F-measure										
	SVM	NB	RF	MLP	LR	BN	AB	DT	KNN	DNN	SVM	NB	RF	MLP	LR	BN	AB	DT	KNN	DNN	
D1	82	84	81	83	82	83	85	81	83	**86**	0.72	0.81	0.85	0.83	0.81	0.83	0.85	0.82	0.83	0.81	**0.87**
D2	80.88	84	87	86	85	82	85	86	88	**89**	0.71	0.83	0.86	0.85	0.84	0.81	0.85	0.83	0.85	0.81	**0.87**
D3	81	83	87	82	85	82	83	84	86	**88**	0.68	0.81	0.80	0.81	0.83	0.83	0.82	0.81	0.83	0.81	**0.84**
D4	78	76	79	76	73	79	75	77	79	**80**	0.62	0.70	0.78	0.74	0.77	0.76	0.76	0.77	0.71	0.76	**0.80**
D5	77.7	76	79	80	81	82	78	75	76	**82.4**	0.67	0.72	0.77	0.74	0.76	0.75	0.74	0.75	0.76	0.77	**0.79**
D6	75	77	76	75	74	80	80	81	83	**84.9**	0.69	0.78	0.75	0.76	0.77	0.77	0.80	0.81	0.82	0.81	**0.86**
D7	80	83	84	86.6	82.7	83	84.8	81.7	81	**85**	0.70	0.80	0.85	0.84	0.87	0.89	0.86	0.86	0.82	0.81	**0.88**
D8	72	75	78	82	84	85	86	87	88	**88.7**	0.67	0.81	0.81	0.88	0.85	0.84	0.83	0.84	0.84	0.88	**0.89**
D9	84	87	82	81	83	**87**	84	85	83	86	0.78	0.81	0.82	0.84	0.83	0.82	**0.85**	0.82	0.81	0.82	0.81
D10	78	83	84	81	82	83	82	82.8	82.7	**85**	0.70	0.82	0.81	0.82	0.84	0.85	0.86	0.87	0.82	0.88	**0.87**
D11	81	80	86	86	84	**87**	81	82	82	83	0.72	0.81	0.82	0.83	0.80	0.82	**0.84**	0.81	0.82	0.81	0.83
D12	78	80	82	86	81	83	88	87	86	**90**	0.70	0.80	0.81	0.82	0.81	0.81	0.82	0.86	0.86	0.84	**0.89**
D13	78	81	81	82	82	87	82	86	88	**89**	0.60	0.78	0.80	0.81	0.80	0.79	0.81	0.82	0.80	0.81	**0.88**
D14	81	80	86	82	85	**87.9**	82	81	86	86	0.67	0.82	0.86	0.81	0.80	0.88	**0.89**	0.83	0.85	0.81	0.82
D15	88	88	86	86	88	**90**	88	89.8	81	82	0.69	0.86	0.82	0.83	0.80	0.81	**0.84**	0.83	0.84	0.86	0.86
D16	71	80	87	86	82	83	81	86	87	**88**	0.72	0.82	0.81	0.80	0.80	0.81	0.83	0.80	0.80	0.81	**0.83**
D17	86	88	86	89	90	91	90	90.1	91	**92**	0.67	0.86	0.82	0.85	0.87	0.88	0.82	0.85	0.88	0.92	**0.93**
D18	87	89	91	90	92.8	91	92.9	91	92	**93**	0.70	0.89	0.85	0.85	0.87	0.84	0.85	0.86	0.88	0.89	**0.90**
D19	86	87	89	91	92	93	92	93	92	**93.7**	0.72	0.89	0.86	0.87	0.88	0.92	0.91	0.92	0.90	0.88	**0.95**
D20	88	86	86	82	84	**88**	82	83	81	82	0.78	0.82	0.86	0.83	0.82	0.85	**0.89**	0.87	0.88	0.84	0.85
D21	78	78	80	81	80	**85.7**	80.7	82	83	84	0.67	0.82	0.85	0.84	0.85	0.87	**0.9**	0.88	0.88	0.87	0.89
D22	83	85	87	88	88	82	82	86	85	**89**	0.68	0.82	0.88	0.87	0.85	0.82	**0.85**	0.85	0.88	0.86	0.89
D23	80	81	80.88	83.76	87.78	87.71	87.9	**90**	89	86	0.68	0.72	0.809	0.846	0.80	0.82	**0.85**	0.83	0.82	0.80	0.82

(continued)

Table 7.8 (continued)

	Accuracy										F-measure										
D24	81	80	82	83	86	**88.9**	88.2	87	87.7	88	0.67	0.80	0.81	0.82	0.86	0.85	**0.89**	0.82	0.82	0.85	0.85
D25	86	88	86	87	87	82	83	86	87	**88.8**	0.77	0.88	0.81	0.86	0.88	0.88	0.89	0.89	0.89	0.88	**0.9**
D26	84	85.7	86.9	86.6	84.8	86.1	84	81.8	82	**89.9**	0.73	0.80	0.83	0.84	0.85	0.83	0.83	0.85	0.81	0.83	**0.89**
D27	82	85	86	81	80.1	81.9	83.6	87	85.8	**87.9**	0.71	0.82	0.86	0.88	0.86	0.83	0.86	0.84	0.81	0.82	**0.89**
D28	82	81	82	82	82.8	**84**	83	82	81	82.9	0.78	0.78	0.82	0.85	0.82	0.81	**0.87**	0.82	0.81	0.82	0.84
D29	78	82	85	86	82	**89**	87	85	84	88	0.77	0.80	0.81	0.84	0.86	0.88	**0.87**	0.84	0.86	0.85	0.88
D30	76	82	84	87	83	81	82	85	81	**87**	0.67	0.72	0.71	0.82	0.85	0.86	0.83	0.85	0.84	0.82	**0.89**

Table 7.9 Accuracy and F-measure using FR3

ID	Accuracy										F-measure									
	SVM	NB	RF	MLP	LR	BN	AB	DT	KNN	DNN	SVM	NB	RF	MLP	LR	BN	AB	DT	KNN	DNN
D1	73.33	75.0	77.37	77.66	78.4	77	78.4	76.6	77.6	78.4	0.79	0.77	0.76	0.79	0.72	0.71	0.79	0.79	0.79	0.79
D2	70	72.08	76.27	72.66	78	76	73	76.6	77.6	83	0.78	0.72	0.74	0.73	0.72	0.76	0.77	0.76	0.76	0.82
D3	82	82.8	83.67	81.06	84	83	82	82.6	81.6	89.7	0.81	0.82	0.82	0.83	0.84	0.83	0.82	0.84	0.83	0.88
D4	72	71.08	72.27	71.60	76	75	72	75.6	79.6	82	0.73	0.70	0.72	0.71	0.70	0.72	0.79	0.78	0.77	0.80
D5	76	81	82	86	82	85	81	86	87	89	0.80	0.84	0.85	0.83	0.86	0.84	0.82	0.84	0.81	0.89
D6	72	72.08	76.27	75.66	78	76	81	86.6	85.6	88	0.78	0.72	0.74	0.73	0.72	0.76	0.80	0.82	0.83	0.89
D7	80	82	86	87	87	85	80	88	87	89	0.80	0.80	0.84	0.83	0.84	0.82	0.86	0.88	0.89	0.90
D8	72	78	79.7	81	83	84	85	88	87	90	0.81	0.80	0.82	0.81	0.82	0.82	0.83	0.83	0.84	0.87
D9	82	87	89	89	90	90.8	93	91	91	92	0.83	0.86	0.84	0.91	0.90	0.92	0.95	0.90	0.90	0.94
D10	75	80	88	89	87	86	85	89	88	90	0.72	0.78	0.76	0.73	0.77	0.75	0.82	0.84	0.83	0.89
D11	78	80	84	84.7	88.7	88	81	85	87	89	0.77	0.76	0.80	0.81	0.84	0.83	0.83	0.84	0.83	0.90
D12	81	86	85	88	87	85	88	82	85	89	0.84	0.83	0.86	0.85	0.84	0.83	0.86	0.86	0.88	0.89
D13	79.9	84.6	87.3	88.7	86.7	87	88	88.2	88.1	89.8	0.78	0.82	0.84	0.83	0.81	0.85	0.83	0.84	0.83	0.89
D14	83.8	86.8	86.3	82	85	87	86	87.6	87.3	89.7	0.80	0.86	0.84	0.83	0.88	0.83	0.84	0.83	0.86	0.88
D15	88	84	88	89	89	89.9	89.8	91.8	90.7	92.4	0.84	0.85	0.86	0.83	0.84	0.85	0.88	0.88	0.89	0.91
D16	75	85	86	85	83	82	80	84	85	88	0.78	0.84	0.83	0.83	0.86	0.84	0.85	0.84	0.83	0.88
D17	86	88	86	89	87.7	88	88.4	87	86.8	89.9	0.82	0.80	0.85	0.83	0.85	0.86	0.88	0.86	0.88	0.89
D18	82	87	89	89.7	88	87	88	86	89	89.9	0.80	0.82	0.85	0.84	0.83	0.85	0.86	0.86	0.86	0.88
D19	73	76	80	81	82	83	84.8	85.8	86.9	88	0.85	0.82	0.85	0.86	0.84	0.81	0.82	0.80	0.82	0.87
D20	76	77	78.6	79.8	80	81.9	88.3	89.6	81	89	0.72	0.76	0.73	0.80	0.78	0.79	0.80	0.81	0.80	0.79
D21	79	78	80	80.9	87.8	88.7	89.8	90	89.7	90.7	0.82	0.81	0.82	0.83	0.84	0.84	0.83	0.81	0.80	0.85
D22	78	82	83	84.7	86	81	82	84	85	88	0.80	0.81	0.83	0.84	0.83	0.84	0.83	0.84	0.85	0.89
D23	78	81	80.88	83.76	88.78	83.71	83.9	89	84	88	0.72	0.79	0.82	0.80	0.81	0.87	0.83	0.82	0.80	0.81

(continued)

Table 7.9 (continued)

	Accuracy										F-measure									
D24	78	80	80.8	82	88	88.2	**89.3**	89	87.7	82	0.78	0.80	0.81	0.83	0.84	0.81	**0.87**	0.82	0.85	0.85
D25	80	82	84	85	88	89	87	89	84	**89.8**	0.80	0.84	0.86	0.87	0.85	0.83	0.82	0.81	0.80	**0.88**
D26	74	78.7	76.9	79.6	80.8	82.1	84	85	85	**88**	0.78	0.80	0.82	0.83	0.85	0.84	0.85	0.86	0.86	**0.87**
D27	80	83	84	81	88	81.9	89.6	87	86.9	**89**	0.78	0.76	0.80	0.82	0.81	0.82	0.83	0.84	0.82	**0.88**
D28	84	85	86	87	89	88	**89.8**	86	81	88	0.78	0.80	0.83	0.84	0.86	0.84	0.81	0.86	0.85	**0.88**
D29	78	80	82	83	81	87	**88.9**	85	84	81.9	0.78	0.80	0.83	0.85	0.87	0.86	0.87	**0.88**	0.81	0.82
D30	76	82	84	83	84	89	89	88	81	**89.8**	0.72	0.80	0.81	0.83	0.84	0.85	0.86	0.83	0.81	**0.88**

Table 7.10 Accuracy and F-measure using FR4

ID	Accuracy										F-measure									
	SVM	NB	RF	MLP	LR	BN	AB	DT	KNN	DNN	SVM	NB	RF	MLP	LR	BN	AB	DT	KNN	DNN
D1	84.33	85.66	87.57	81.66	**88.4**	81	**88.4**	80.6	80.6	**88.4**	**0.89**	0.87	0.86	**0.89**	0.82	0.81	**0.89**	**0.89**	0.79	**0.89**
D2	78.8	78.08	79.27	79.66	80	81	83	86.6	87	**88**	0.75	0.78	0.79	0.83	0.82	0.86	0.87	0.86	0.86	**0.89**
D3	82	84.8	87.67	85.06	87	82	84	83.6	84.6	**86.7**	0.77	0.80	0.81	0.84	0.85	0.86	0.84	0.85	0.82	**0.85**
D4	76	77.08	78.27	76.60	78	80	81	79.6	80.6	**81.9**	0.70	0.73	0.76	0.77	0.78	0.79	0.80	0.79	0.76	**0.81**
D5	79	83	86	88	85	83	84	83	84	**88.9**	0.79	0.82	0.86	0.81	0.84	0.82	0.80	0.81	0.84	**0.85**
D6	78	78.08	79.27	79.66	79	80	82	85.5	82.9	**87**	0.78	0.75	0.74	0.73	0.71	0.76	0.80	0.82	0.82	**0.84**
D7	78.9	82	83	84	85	86	88	89	87	**89.8**	0.82	0.81	0.83	0.84	0.85	0.81	0.86	0.87	0.80	**0.89**
D8	72	78	79.7	82	85.8	84	85	88	87	**89**	0.81	0.80	0.82	0.81	0.82	0.82	0.83	0.83	0.84	**0.87**
D9	82	87.7	89	89	90	90.8	**93.7**	91	91	93	0.83	0.86	0.84	0.91	0.90	0.92	**0.95**	0.90	0.90	0.94
D10	75	80	88	89.9	87.8	86	85	89	88	**89**	0.71	0.76	0.76	0.73	0.77	0.75	0.82	0.84	0.83	**0.85**
D11	78	80	84	85.7	86.1	82	81	85	87	**88.7**	0.72	0.74	0.80	0.81	0.84	0.83	0.81	0.82	0.83	**0.86**
D12	80.8	86.8	85.6	88.6	87.4	85.1	88.5	82.3	85.1	**88.9**	0.74	0.83	0.80	0.85	0.81	0.82	0.85	0.83	0.87	**0.88**
D13	79.9	82.6	87	86.7	86.7	87	81	82.2	87.1	**87.8**	0.76	0.81	0.83	0.85	0.80	0.85	0.81	0.83	0.80	**0.86**
D14	83.8	86.8	86.3	82	85	87	86	84.6	85.3	**86.7**	0.82	0.84	0.84	0.83	0.88	0.83	0.84	0.83	0.86	**0.87**
D15	85	86	88.8	89.5	88.7	88.9	88.8	81.8	80.7	**89.8**	0.81	0.86	0.85	0.86	0.85	0.80	0.84	0.88	0.87	**0.89**
D16	75	85	86	85	83	82	80	84	85	**88**	0.78	0.84	0.83	0.83	0.86	0.84	0.85	0.84	0.83	**0.88**
D17	86	88	86	82	86.7	88	86.4	87	82.8	**89**	0.82	0.80	0.85	0.83	0.85	0.86	0.88	0.86	0.88	**0.89**
D18	80	81	86	82.7	83	85	82	86	81	**86.9**	0.80	0.82	0.85	0.84	0.80	0.81	0.83	0.82	0.84	**0.85**
D19	73	76	82	83	85	84	85.8	86.8	87.9	**88.8**	0.81	0.82	0.83	0.85	0.82	0.80	0.82	0.80	0.82	**0.86**
D20	76	77	78.6	79.8	80.8	81.9	84.3	**86.6**	81	89	0.72	0.74	0.72	0.78	0.76	0.77	0.79	**0.80**	0.78	0.79
D21	79	78	80	82.9	87.8	87.7	86.8	87.8	83.7	**89.7**	0.81	0.83	0.85	0.87	0.85	0.87	0.86	0.87	0.82	**0.88**
D22	78	82	83	84.7	85	81	82	84	85	**86**	0.80	0.81	0.83	0.84	0.83	0.84	0.83	0.84	0.85	**0.89**
D23	78	81	80.88	83.76	88	83.71	83.9	**89.2**	84	88	0.72	0.79	0.82	0.80	0.81	**0.87**	0.83	0.82	0.80	0.81

(continued)

Table 7.10 (continued)

	Accuracy										F-measure									
D24	78	80	80.8	82	86	88.2	**88.6**	89	85.7	81	0.78	0.80	0.81	0.83	0.81	0.83	**0.86**	0.82	0.84	0.83
D25	80	82	84	85	82	84	83	87	84	**87.8**	0.81	0.82	0.84	0.85	0.86	0.85	0.83	0.80	0.82	**0.87**
D26	74	78.7	77.9	79.6	80.8	82.1	84	85	82	**87**	0.78	0.81	0.82	0.82	0.84	0.84	0.85	0.81	0.82	**0.85**
D27	80	83	84	81	82	80.9	89.2	81	85.9	**86**	0.78	0.76	0.80	0.82	0.81	0.82	0.81	0.84	0.82	**0.86**
D28	84	85	86	84	83	85	87.8	**89**	81	88	0.78	0.80	0.83	0.81	0.84	0.85	0.81	0.80	0.84	**0.86**
D29	78	80	82	83	81	87	**87.9**	85	84	83.7	0.78	0.82	0.81	0.83	0.85	0.81	0.83	**0.87**	0.83	0.84
D30	76	81	82	83	84.7	89.6	89.2	88	84	**89.8**	0.72	0.80	0.81	0.83	0.84	0.85	0.82	0.81	0.84	**0.89**

Table 7.11 Accuracy and F-measure using FR5

ID	Accuracy										F-measure									
	SVM	NB	RF	MLP	LR	BN	AB	DT	KNN	DNN	SVM	NB	RF	MLP	LR	BN	AB	DT	KNN	DNN
D1	82	84	86	86	82	83	88	87	84	**89**	0.80	0.83	0.84	0.85	0.86	0.87	0.82	0.88	0.81	**0.89**
D2	81.88	83	85	84	83	86	87	89	89	**90**	0.80	0.85	0.83	0.82	0.80	0.84	0.80	0.83	0.85	**0.86**
D3	81	84	87	89	86	82	84	85	88.9	**89.9**	0.87	0.86	0.85	0.83	0.85	0.86	0.85	0.84	0.85	**0.88**
D4	88	86	84	85	83	88	85	87	89	**90.9**	0.82	0.87	0.85	0.86	0.86	0.86	0.88	0.81	0.88	**0.89**
D5	83	80	81	86	88	82	83	84	85	**88.7**	0.80	0.81	0.82	0.83	0.83	0.84	0.85	0.86	0.82	**0.89**
D6	85	87	86	85	84	88	89	90.8	90	**91**	0.88	0.85	0.86	0.87	0.87	0.85	0.88	0.89	0.89	**0.90**
D7	85	88	89	89.6	88.7	86	86.8	89.7	90	**90.8**	0.86	0.85	0.84	0.87	0.89	0.86	0.87	0.88	0.81	**0.89**
D8	77	78	78	82	84	85	86	87	87.9	**90.9**	0.81	0.81	0.88	0.85	0.84	0.83	0.84	0.84	0.87	**0.88**
D9	84	87	90	88	83	84	**89**	**89**	87	**89**	0.89	0.82	0.84	0.83	0.82	0.89	**0.90**	0.81	0.82	**0.89**
D10	78	86	86	88	82	89	89	87.8	86.7	**89.9**	0.82	0.81	0.88	0.82	0.85	0.88	0.81	0.82	0.88	**0.89**
D11	88	88	86	86	89	89	80	86	88	**91**	0.87	0.86	0.87	0.86	0.87	0.85	0.84	0.82	0.85	**0.89**
D12	82	88	82	86	81	83	87	88	89.1	**89.8**	0.81	0.80	0.82	0.81	0.83	0.82	0.86	0.86	0.84	**0.88**
D13	78	81	80	82	82	87	82	86	88	**89.6**	0.78	0.80	0.81	0.80	0.79	0.81	0.83	0.85	0.85	**0.89**
D14	81	80	86	82	85	82	81	86	86	**87.9**	0.82	0.86	0.81	0.80	0.81	0.83	0.83	0.84	0.83	**0.86**
D15	88	88	86	86	88	88	89.8	88.7	82.9	**88.9**	0.86	0.78	0.80	0.86	0.85	0.83	0.81	0.82	0.84	**0.87**
D16	71	80	87	86	82	83.8	81	86.9	87.8	**88.1**	0.82	0.81	0.81	0.80	0.82	0.83	0.80	0.80	0.81	**0.84**
D17	86	88	86	89	89	88	83	86	81	**89.7**	0.86	0.82	0.85	0.82	0.78	0.72	0.80	0.78	0.88	**0.89**
D18	84	86	81	84	82.8	85	82.9	87	86	**89**	0.81	0.85	0.85	0.87	0.82	0.85	0.86	0.84	0.88	**0.9**
D19	83	86	88	88	89	87	86	87	89	**89.7**	0.81	0.81	0.84	0.84	0.88	0.81	0.82	0.80	0.88	**0.88**
D20	88	86	86	88	89	85	**88**	85	81	82	0.81	0.82	0.81	0.84	0.83	0.85	0.86	**0.87**	0.82	0.83
D21	78	78	80	81	80	81.7	82.7	83.4	84.2	**88.7**	0.82	0.85	0.84	0.85	0.87	0.80	0.81	0.87	0.86	**0.88**
D22	83	82	85	86	86	83	81	82	85	**86.9**	0.80	0.83	0.85	0.81	0.80	0.82	0.82	0.85	0.85	**0.86**
D23	80	81	80.88	82.76	81.78	85.71	86.9	**89.9**	81	80.1	0.72	0.80	0.84	0.80	0.82	**0.85**	0.83	0.82	0.80	0.82

(continued)

Table 7.11 (continued)

	Accuracy										F-measure									
D24	80	81	81	82	83	86.4	**88.3**	85	86.7	82	0.72	0.80	0.81	0.82	0.83	0.80	**0.84**	0.81	0.82	0.83
D25	86	84	84	84	87	85	83	86	87	**88.1**	0.88	0.81	0.86	0.88	0.88	0.82	0.85	0.86	0.88	**0.89**
D26	81	82.7	83.9	84.6	84.8	86.1	84	87	81	**87.1**	0.82	0.83	0.81	0.82	0.83	0.81	0.82	0.86	0.85	**0.89**
D27	82	82	84	80	80.1	82.9	83.6	86.8	85.8	**87.9**	0.82	0.82	0.84	0.84	0.85	0.87	0.85	0.86	0.87	**0.88**
D28	82	86	89	86	82.8	81.1	**86.5**	82	81.9	81.9	0.78	0.80	0.82	0.81	0.84	0.83	**0.85**	0.80	0.81	0.83
D29	78	82	85	86	82	87	**88.9**	85	84	88.4	0.80	0.81	0.84	0.86	0.88	0.87	0.86	**0.87**	0.85	0.86
D30	76	82	84	87	89	81	82	85	81	**89.6**	0.72	0.71	0.82	0.85	0.86	0.87	0.85	0.87	0.87	**0.88**

Table 7.12 Accuracy and F-measure using FR6

ID	Accuracy										F-measure									
	SVM	NB	RF	MLP	LR	BN	AB	DT	KNN	DNN	SVM	NB	RF	MLP	LR	BN	AB	DT	KNN	DNN
D1	81	84	86	86	82	83	85	86	83	**89.8**	0.81	0.85	0.83	0.81	0.83	0.85	0.82	0.87	0.81	**0.89**
D2	80.88	84	87	86	85	82	85	86	89	**91.8**	0.82	0.86	0.85	0.84	0.81	0.85	0.83	0.85	0.81	**0.87**
D3	81	84	87	82	85	83	81	84	89	**90.8**	0.87	0.86	0.85	0.83	0.85	0.86	0.85	0.84	0.87	**0.89**
D4	78	81	89	86	83	89	85	87	89	**90.7**	0.80	0.88	0.84	0.87	0.86	0.86	0.87	0.81	0.86	**0.88**
D5	81	83	80	81	86	87	82	83	85	**89.8**	0.80	0.81	0.82	0.83	0.83	0.84	0.85	0.86	0.87	**0.90**
D6	85	87	86	85	84	88	89	92	94	**96.7**	0.88	0.85	0.86	0.87	0.87	0.85	0.88	0.87	0.88	**0.90**
D7	81	85	88	89	89.6	88.7	86	86.8	89.7	**93.8**	0.86	0.85	0.84	0.87	0.89	0.86	0.87	0.82	0.81	**0.89**
D8	78	75	78	82	84	85	86	87	88	**91**	0.81	0.81	0.88	0.85	0.84	0.83	0.84	0.84	0.88	**0.89**
D9	84	87	92	91	83	84	**96**	95	93	86	0.89	0.92	0.94	0.93	0.92	0.96	**0.99**	0.91	0.92	0.91
D10	78	86	88	89	82	89	89	89.8	89.7	**97**	0.82	0.81	0.88	0.86	0.87	0.85	0.88	0.82	0.88	**0.96**
D11	88	88	86	86	89	89	80	86	88	**98**	0.87	0.86	0.86	0.85	0.87	0.85	0.84	0.82	0.85	**0.93**
D12	82	88	82	86	81	83	88	87	89	**90**	0.80	0.81	0.82	0.81	0.81	0.82	0.86	0.86	0.84	**0.89**
D13	78	81	81	82	82	87	82	86	88	**89.8**	0.78	0.80	0.81	0.80	0.79	0.81	0.82	0.80	0.81	**0.88**
D14	81	80	86	82	85	82	81	86	86	**90.9**	0.82	0.86	0.81	0.80	0.88	0.83	0.85	0.81	0.82	**0.89**
D15	88	88	86	86	88	88	89.8	91	92	**97**	0.86	0.88	0.83	0.86	0.85	0.83	0.84	0.86	0.86	**0.94**
D16	71	80	87	86	82	83	81	86	87	**88**	0.82	0.81	0.80	0.80	0.81	0.83	0.80	0.80	0.81	**0.83**
D17	86	88	86	89	90	92	93	96	91	**98**	0.86	0.82	0.85	0.87	0.88	0.82	0.85	0.88	0.98	**1**
D18	87	89	91	90	92.8	91	92.9	95	96	**98.9**	0.89	0.85	0.85	0.87	0.84	0.85	0.86	0.88	0.90	**0.93**
D19	86	87	89	91	92	93	92	95	96	**97**	0.89	0.86	0.87	0.88	0.92	0.91	0.92	0.90	0.88	**0.95**
D20	88	86	86	89	90	92	93	**95**	96	92	0.82	0.86	0.83	0.82	0.85	0.87	0.88	**0.89**	0.84	0.85
D21	78	78	80	81	80	80.7	82	83	84	**85.7**	0.82	0.85	0.84	0.85	0.87	0.88	0.88	0.87	0.89	**0.9**
D22	83	85	87	88	88	90	92	96	95	**98**	0.82	0.88	0.87	0.85	0.82	0.85	0.85	0.88	0.89	**0.91**
D23	80	81	80.88	83.76	87.78	87.71	87.9	**91**	90	90.1	0.72	0.809	0.846	0.80	0.82	**0.85**	0.83	0.82	0.80	0.82

(continued)

Table 7.12 (continued)

	Accuracy										F-measure									
D24	81	80	82	83	89	89.2	**91.3**	90	89.7	88	0.80	0.81	0.82	0.86	0.85	0.82	**0.89**	0.82	0.85	0.85
D25	86	88	86	89	97	92	93	96	97	**98.8**	0.88	0.81	0.86	0.88	0.88	0.89	0.89	0.89	0.88	**1**
D26	84	88.7	86.9	89.6	94.8	96.1	94	97	95	**97.9**	0.82	0.83	0.84	0.85	0.86	0.86	0.85	0.88	0.86	**0.92**
D27	82	85	86	81	90.1	91.9	93.6	97	95.8	**97.9**	0.82	0.86	0.88	0.89	0.89	0.87	0.85	0.86	0.87	**0.91**
D28	82	86	89	86	92	91	**98**	86	81	91	0.78	0.82	0.85	0.82	0.85	0.86	0.89	0.88	0.85	0.88
D29	78	82	85	86	82	87	**92**	85	84	89	0.80	0.81	0.84	0.86	0.88	0.87	0.86	**0.87**	0.85	0.88
D30	76	82	84	87	89	91	92	95	91	98	0.72	0.71	0.82	0.85	0.86	0.87	0.85	0.87	0.87	**1**

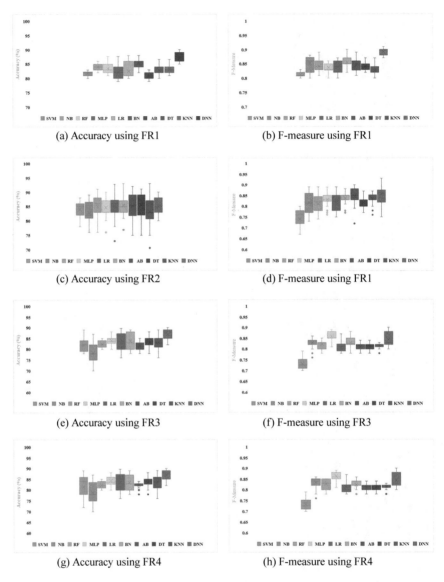

Fig. 7.4 Box-plot diagrams of F-measure and Accuracy

7.9.3.1 Feature Selection Techniques

In the present work, six distinct sets of the features have been recognized as input to build a model over thirty distinct categories of Android apps. Ten different classifiers have been recognized to build a detection model by examining F-measure and Accuracy. In accordance with feature selection approach, two sets are utilized, each

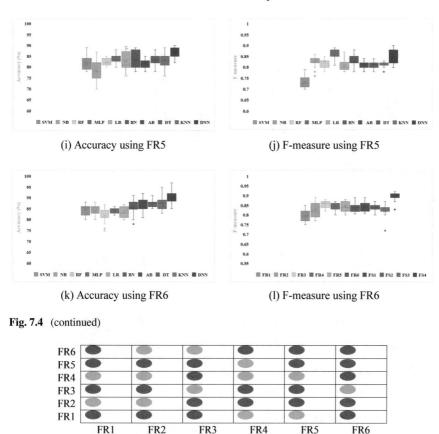

(i) Accuracy using FR5

(j) F-measure using FR5

(k) Accuracy using FR6

(l) F-measure using FR6

Fig. 7.4 (continued)

Fig. 7.5 t-test analysis (p-value) for feature ranking techniques

with 300 points (10 classifier × 30 data sets). Here, *t*-test among distinct feature selection approaches are conducted and as matched the equivalent P-value to evaluate statistical importance. Figure 7.5 demonstrates the outcome of the *t*-test study. For simplicity, the P-values are represented by utilizing two distinct symbols such as (green circle) P-value > 0.05 (no relevance difference) and (red circle) <= 0.05 (relevance difference). On the basis of Fig. 7.5 it is seen that most of the cells have green circles; it seems that it does not significantly differentiate among applied feature selection approaches. This means, FR6 selected set of features using PCA yield better results as matched to other approaches.

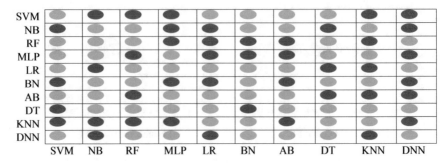

Fig. 7.6 t-test analysis (p-value) for classification methods

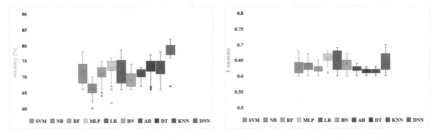

(a) Accuracy using all extracted features   (b) F-measure using all extracted features

Fig. 7.7 Box-plot diagrams of F-measure and accuracy

7.9.3.2   Classification Methods

In the present chapter, ten distinct classification models have been recognized to build a model to detect either the app is benign or malware. This work, uses six distinct set of features (six feature ranking approach) of thirty distinct Android categories apps with two performance parameters F-measure and Accuracy, so for each detection approach a total number of 180 data point (6 feature ranking approach × 30 data sets of Android apps). Figure 7.6 demonstrates the outcomes of t-test analysis. On the basis of Fig. 7.6, it is clear that, there is a significant difference between two techniques because p-value is smaller than 0.05 (represent circle with green colour). But upon judging the value, DNN gives better outcomes matched to other classifiers.

7.9.3.3   Classifier with All Selected Features

In this section, ten different classifiers have been only considered to develop a model to detect either the app is benign or malware without selected features (means using all the extracted features). Figure 7.7 demonstrates the box-plot diagram of F-measure and Accuracy by using ten different classifiers with all extracted features.

Table 7.13 Comparison with AV scanners

AV scanners	Accuracy (%)
AV1 (Panda Free Antivirus)	86
AV2 (Avast Free Antivirus)	93
AV3 (Adaware Antivirus Free)	39
AV4 (Comodo Antivirus 10)	88
AV5 (AVG AntiVirus FREE)	92
Proposed approach (FR6 + DNN)	97.8

Table 7.14 Comparison with previously used classifiers

Name of the machine learning classifier	Averaged accuracy (%)
SimpleLogistic [20]	84.08
BayesNet K2 [20]	82
BayesNet TAN [20]	68.51
RandomTree [20]	83.32
PerbDroid (our proposed framework)	97.8

7.9.4 Evaluation of PerbDroid Using Detection Framework

7.9.4.1 Comparison with AV Scanners

To evaluate the performance of our proposed framework, we use five different anti-virus scanners available in the market. These anti-virus scanners work on signature based approach. For this, we download free 1000 .apk files from different sources and use AV scanners to detect malware from them. The accuracy of different AV scanners mentioned in Table 7.13. From Table 7.13, it is seen that our proposed approach (FR6 + DNN) able to detect 97.8% malware apps where as various anti-virus scanners are able to detect only 39 to 93% malware apps.

7.9.4.2 Comparison with Previously Used Classifiers

In this subsection, we compare the performance of PerbDroid with existing used classifiers. Table 7.14 show us the comparison with existing classifiers. From Table 7.14, it is seen that our proposed framework is given average accuracy of 97.8% when applied on thirty different categories of Android app which is more than average accuracy of previously used classifiers in the literature.

7.9.5 Experimental Finding

This part of the chapter contains, the overall finding of the experimental works. The experimental work was conducted for thirty different categories of Android apps by selected features with the help of six different feature ranking techniques. Further, the selected features are trained with ten different classifiers and the performance of the respective techniques measured by using two performance parameters i.e., F-measure and Accuracy. Based on this experimental work performed, this chapter helps to answer the following research questions.

RQ1: In this work, ten distinct classifiers have been recognized to build a model to detect either the app is malware or benign. On the basis of Tables 7.7, 7.8, 7.9, 7.10, 7.11 and 7.12, it can be inferred that model build using DNN by considering selected set of features by utilizing FR6 as input gives better results as matched to others.

RQ2: To respond RQ2, Fig. 7.4 were examined, it is seen that the performance of the feature selection approaches is changed with the distinct classification approaches used. It is seen that selection of classification approach to build a detection model to detect either the app is benign or malware is influenced by the feature selection approaches.

RQ3: In this work, six distinct kinds of feature ranking approaches are recognized to discover the reduced sub-set of features. On the basis of t-test study, it is clear that feature selection by utilizing FR6 i.e., PCA approach produces the best outcomes as compared to others.

RQ4: To answer RQ4, Figs. 7.4 and 7.7 were analyzed, by using six different features ranking methods our developed model is more capable to detect malware rather than considering all extracted features.

7.9.6 Conclusion and Future Work

This chapter highlighted on designing a malware detection framework for determining the effectiveness of the build malware detection model which is developed by utilizing set of features. For this, thirty different set of features are utilized to build a model by using ten different classifier. The execution process was conducted by taking help of thirty different categories of Android app. The experiments performed and outcomes generated on *MATLAB* environment.

Our submissions are the following:

- Our research outcomes propose that, it is possible to construct a malware detection model based on a selected set of features. Further, it is observed that this malware detection model build using this recognized set of features is able to detect benign and malware apps with better accuracy and reduced value of misclassification errors.

- On the basis of research findings, we seen that even after reducing 50% (average) of the available number of features, the build malware detection models were not adversely affected; actually, in the majority of cases the outcomes were better.

In this chapter, developed models for malware detection only detects that either the app is benign or malware. Further, study can be extended to identify how many number of features are required to judge that the app belong to which category (i.e., benign or malware). Further, this study can be replicated over other framework which use soft computing models to obtain better accuracy for malware detection.

References

1. https://file.gdatasoftware.com/web/en/documents/whitepaper/G_DATA_Mobile_Malware_Report_H1_2016_EN.pdf
2. Allix, K., Bissyandé, T.F., Jérome, Q., Klein, J., Le Traon, Y.: Empirical assessment of machine learning-based malware detectors for android. Empir. Softw. Eng. **21**(1), 183–211 (2016)
3. Narudin, F.A., Feizollah, A., Anuar, N.B., Gani, A.: Evaluation of machine learning classifiers for mobile malware detection. Soft Comput. **20**(1), 343–357 (2016)
4. Afifi, F., Anuar, N.B., Shamshirband, S., Choo, K.K.R.: DyHAP: dynamic hybrid ANFIS-PSO approach for predicting mobile malware. PloS One **11**(9), e0162627 (2016)
5. Ab Razak, M.F., Anuar, N.B., Salleh, R., Firdaus, A.: The rise of "malware": bibliometric analysis of malware study. J. Netw. Comput. Apps **75**, 58–76 (2016)
6. Aafer, Y., Du, W., Yin, H.: Droidapiminer: mining api-level features for robust malware detection in android. In: International Conference on Security and Privacy in Communication Systems, pp. 86–103. Springer, Cham (2013)
7. Talha, K.A., Alper, D.I., Aydin, C.: APK auditor: permission-based android malware detection system. Digit. Investig. **13**, 1–14 (2015)
8. Suarez-Tangil, G., Tapiador, J.E., Peris-Lopez, P., Blasco, J.: Dendroid: a text mining approach to analyzing and classifying code structures in android malware families. Expert. Syst. Appl. **41**(4), 1104–1117 (2014)
9. Firdaus, A., Anuar, N.B., Ab Razak, M.F., Sangaiah, A.K.: Bio-inspired computational paradigm for feature investigation and malware detection: interactive analytics. Multimed. Tools Appl. 1–37 (2017)
10. Yuan, Z., Lu, Y., Xue, Y.: Droiddetector: android malware characterization and detection using deep learning. Tsinghua Sci. Technol. **21**(1), 114–123 (2016)
11. Wu, D.J., Mao, C.H., Wei, T.E., Lee, H.M., Wu, K.P.: Droidmat: android malware detection through manifest and api calls tracing. In: 2012 Seventh Asia Joint Conference on Information Security, pp. 62–69. IEEE (2012)
12. Faruki, P., Ganmoor, V., Laxmi, V., Gaur, M.S., Bharmal, A.: AndroSimilar: robust statistical feature signature for Android malware detection. In: Proceedings of the 6th International Conference on Security of Information and Networks, pp. 152–159. ACM (2013)
13. Wu, S., Wang, P., Li, X., Zhang, Y.: Effective detection of android malware based on the usage of data flow APIs and machine learning. Inf. Softw. Technol. **75**, 17–25 (2016)
14. Qiao, M., Sung, A. H., Liu, Q.: Merging permission and API features for android malware detection. In: 2016 5th IIAI International Congress on Advanced Applied Informatics (IIAI-AAI), pp. 566–571. IEEE (2016)
15. Wang, S., Yan, Q., Chen, Z., Yang, B., Zhao, C., Conti, M.: Detecting android malware leveraging text semantics of network flows. IEEE Trans. Inf. Forensics Secur. **13**(5), 1096–1109 (2017)

16. Zhu, H.J., You, Z.H., Zhu, Z.X., Shi, W.L., Chen, X., Cheng, L.: DroidDet: effective and robust detection of android malware using static analysis along with rotation forest model. Neurocomputing **272**, 638–646 (2018)

17. Vinayakumar, R., Soman, K.P., Poornachandran, P., Sachin Kumar, S.: Detecting android malware using long short-term memory (LSTM). J. Intell. Fuzzy Syst. **34**(3), 1277–1288 (2018)

18. Shen, F., Del Vecchio, J., Mohaisen, A., Ko, S.Y., Ziarek, L.: Android malware detection using complex-flows. IEEE Trans. Mob. Comput. (2018)

19. Huang, C.Y., Tsai, Y.T., Hsu, C.H.: Performance evaluation on permission-based detection for android malware. In: Advances in Intelligent Systems and Applications, vol. 2, pp. 111–120. Springer, Berlin, Heidelberg (2013)

20. Sanz, B., Santos, I., Laorden, C., Ugarte-Pedrero, X., Bringas, P.G., Ãlvarez, G.: Puma: Permission usage to detect malware in android. In: International Joint Conference CISISâ12-ICEUTE 12-SOCO 12 Special Sessions, pp. 289–298. Springer, Berlin, Heidelberg (2013)

21. Shin, W., Kiyomoto, S., Fukushima, K., Tanaka, T.: Towards formal analysis of the permission-based security model for android. In: Fifth International Conference on Wireless and Mobile Communications, 2009. ICWMC'09, pp. 87–92. IEEE (2009)

22. Tang, W., Jin, G., He, J., Jiang, X.: Extending android security enforcement with a security distance model. In: 2011 International Conference on Internet Technology and Applications (iTAP), pp. 1–4. IEEE (2011)

23. Enck, W., Ongtang, M., McDaniel, P.: On lightweight mobile phone application certification. In: Proceedings of the 16th ACM conference on Computer and Communications Security, pp. 235–245. ACM (2009)

24. Burguera, I., Zurutuza, U., Nadjm-Tehrani, S.: Crowdroid: behavior-based malware detection system for android. In: Proceedings of the 1st ACM Workshop on Security and Privacy in Smartphones and Mobile Devices, pp. 15–26. ACM (2011)

25. Shabtai, A., Kanonov, U., Elovici, Y., Glezer, C., Weiss, Y.: "Andromaly": a behavioral malware detection framework for android devices. J. Intell. Inf. Syst. **38**(1), 161–190 (2012)

26. Huang, W., Dong, Y., Milanova, A., Dolby, J.: Scalable and precise taint analysis for android. In: Proceedings of the 2015 International Symposium on Software Testing and Analysis, pp. 106–117. ACM (2015)

27. Mas'ud, M.Z., Sahib, S., Abdollah, M.F., Selamat, S.R., Yusof, R.: Analysis of features selection and machine learning classifier in android malware detection. In: 2014 International Conference on Information Science & Applications (ICISA), pp. 1–5. IEEE (2014)

28. Narayanan, A., Chandramohan, M., Chen, L., Liu, Y.: A multi-view context-aware approach to android malware detection and malicious code localization. Empir. Softw. Eng. 1–53 (2018)

29. Azmoodeh, A., Dehghantanha, A., Choo, K.K.R.: Robust malware detection for internet of (battlefield) things devices using deep eigenspace learning. IEEE Trans. Sustain. Comput. (2018)

30. Kadir, A.F.A., Stakhanova, N., Ghorbani, A.A.: Android botnets: What urls are telling us. In: International Conference on Network and System Security, pp. 78-91. Springer, Cham (2015)

31. Zhou, Y., Jiang, X.: Dissecting android malware: characterization and evolution. In: 2012 IEEE Symposium on Security and Privacy, pp. 95–109. IEEE (2012)

32. Plackett, R.L.: Karl Pearson and the chi-squared test. Int. Stat. Rev./Revue Internationale de Statistique 59–72 (1983)

33. Novakovic, J.: The impact of feature selection on the accuracy of naïve bayes classifier. In: 18th Telecommunications Forum TELFOR, vol. 2, pp. 1113–1116 (2010)

34. Camargo Cruz, A.E., Ochimizu, K.: Towards logistic regression models for predicting fault-prone code across software projects. In: Proceedings of the 2009 3rd International Symposium on Empirical Software Engineering and Measurement, pp. 460–463. IEEE Computer Society (2009)

35. Wang, W., Wang, X., Feng, D., Liu, J., Han, Z., Zhang, X.: Exploring permission-induced risk in android applications for malicious application detection. IEEE Trans. Inf. Forensics Secur. **9**(11), 1869–1882 (2014)

36. Arp, D., Spreitzenbarth, M., Hubner, M., Gascon, H., Rieck, K., Siemens, C.E.R.T.: Drebin: effective and explainable detection of android malware in your pocket. In: Ndss, vol. 14, pp. 23–26 (2014)
37. Yuan, Z., Lu, Y., Wang, Z., Xue, Y.: Droid-sec: deep learning in android malware detection. In: ACM SIGCOMM Computer Communication Review, vol. 44, no. 4, pp. 371–372. ACM (2014)

Chapter 8
A Study on Application of Soft Computing Techniques for Software Effort Estimation

Sripada Rama Sree and Chatla Prasada Rao

Abstract Software is everywhere. Now-a-day' s software plays an indispensable role in all the fields like Education, Medical, Insurance, Marketing, Stock Exchange etc. The major goal of software organization is to achieve the Win-Win condition. As per the Standish Group Chaos Survey, only 30–40% of the software projects are successful. One of the main reasons for failure of the software projects is inaccurate estimations of the cost and schedule. In the conventional software development Algorithmic and Expert Based techniques are used to predict the effort, duration and cost of the software project. But they are not providing accurate estimations of the software effort. Most recently, the industries and researchers are adopting Soft Computing techniques to estimate the software effort accurately in the early stage of software development. In this chapter, the Soft Computing techniques for effort estimation using Fuzzy Logic, Neural Networks (NN), Adaptive Neuro Fuzzy Inference System (ANFIS), Random Forest and Support Vector Machines (SVM) are introduced. Soft Computing models are developed using Fuzzy Logic, Neural Networks and ANFIS using the NASA93 and Desharnais Datasets. Comparing these models using different evaluation criteria, it is observed that the Adaptive Neuro Fuzzy Inference System produced better effort estimates.

8.1 Introduction

Most of the human activities and services today depends partially or completely on Software Systems. The dependence of human life on software systems is increasing rapidly. Most of the surveys are stating that 60–90% manual activities are transformed to automation using software systems. In the conventional days some products and services had been designed and developed by hardware techniques and now these

S. R. Sree (✉) · C. P. Rao
Department of CSE, Aditya Engineering College, Surampalem, India
e-mail: ramasree_s@aec.edu.in

C. P. Rao
e-mail: prasadarao.chatla@aec.edu.in

© Springer Nature Switzerland AG 2020
J. Singh et al. (eds.), *A Journey Towards Bio-inspired Techniques in Software Engineering*,
Intelligent Systems Reference Library 185,
https://doi.org/10.1007/978-3-030-40928-9_8

have been realized by software systems. However, few commercial products and customized services are only feasible for providing solutions through software systems. Consequently, the size of the software and its complexity in various fields has been escalated.

The increasing size and complexity of the software system involves reasonable cost, well defined functionality, time to market, performance requirements etc. for domains like Marketing, Insurance, Hospital, Education, Manufacturing etc. Applications are developed and supported by different software. These software systems must be very fast, cheaper, reliable, intelligent, dependable, flexible and so on. This required list is quite long and will never end. So, software systems must be cheaper and must be delivered to market even before the competitors think about something similar. Barry Boehm states that the "software project is succeeded iff the software must be delivered with a given budget, before a given deadline, all the features of software system are with high quality and the software must be usable [1].

Project management and Software effort estimations are key to success of software development [1]. Software Project Management (SPM) plays a key role in the software development for the successful completion of the projects. The major discriminator for success of a project is good project management rather than a good technology. Irrespective of technology being used, poor project management leads to failure of the project and good project Management leads to success of the software project. The major objective of SPM is to ensure that the software is delivered on time, within budget while meeting all the specified customer needs. The activities of Software Project Management include Project planning, Project budgeting, Scheduling, Risk analysis, Project Staffing and others. As part of the planning, the scope document is to be determined, Business case is to be established and Work Break down Structure (WBS) is to be developed. WBS [2] will help in developing the business case through estimating the schedule and budget of the software project can be done. Business case is a contract between the customer and organizations on budget, schedules, and other deliverables to the customer. The basic questions while doing the estimations are [3]

- What is the total SIZE of the project?
- How much effort (person-months/person-hours) is required to develop the project?
- How long time (Schedule/Calender Time) is needed for each and every activity of the project?
- What is the total budget for the project?
- Identify any other value-added factors which influence the projects.

In the process of estimating the project, the size of the project is to be determined first. Based on the size, the effort required for developing the project is estimated. The development cost and schedule can then be derived from effort. The level of estimation will be refining from phase to phase of software development. In the initial phase of the project, cost, time, ROI (Return on Investment) and other resources are estimated. After beginning of the project, the total cost, time etc. are refined and reconfirmed.

The project management team need to do a lot of exercise for developing and determining the project scope and business case of the software projects in the early stage of the software development (SD). The accurate and early stage estimations lead to the success of the project and otherwise the projects may be failed, challenged or cancelled. Inaccurate estimations will never achieve the Win-Win condition in the software organizations. The foremost reason for the failure of software projects is inaccurate estimation of the software cost and schedule. So, the project management will be the major discriminator to get success or failure of the software projects. Estimates are the basis for bidding, planning, budgeting and scheduling of the software development. Effort predictions can be done with different parameters like Lines of Code, Function Points, Object Points, Use Case Points, Story points etc.

8.2   Software Size Metrics

For estimating software effort, the major input is the Software Size. By far, most effort estimation models for development of the software projects or maintenance of software projects use the software size as the main input driver. Different Techniques, methodologies and tools available in software organizations to measure the size of the software. The most widely used metrics for software sizing available in the industry are Lines Of Code (LOC), Function Point Analysis (FPA), Use Case Points (UCP), Story Points etc. Each and every metric has its own advantages and disadvantages and these metrics may be recommended to estimate different types of software projects.

8.2.1   Lines Of Code (LOC)

The common size metric in software development is Lines Of Code (LOC) [4]. It is generally utilized and acknowledged by the majority of the organization. It is one of the efficient metrics to measure the software size in earlier days. It could be easily measured when the coding activity is completed in the software development life cycle (SDLC). LOC directly relates to the end product. It is purely from the developer's end where it is actually developed or coded. LOC has its own disadvantages too. It is very difficult to measure LOC in the early stage of SDLC. LOC can be varying with programming languages, design methodologies, test cases generated, programmer's abilities, available environments etc. There are no standards in industry to measure the LOC. LOC is the most usable software size metric in Software Effort Estimation (SEE), Software Cost Estimation (SCE), Software Schedule Estimation (SSE) or Software Maintenance Estimations (SME) etc. LOC is a basic input parameter for estimation of software effort in algorithmic models. The standard algorithmic model—Constructive Cost Model (COCOMO) is used from the last few decades for estimations but these estimations are not so accurate.

8.2.2   Function Point Analysis (FPA)

In the mid 1970's, Allan Albercht while working in IBM, faced a problem to determine the software size in the earlier days as opposed to LOC. He introduced the Function Point Analysis (FPA) [5] as a form of sizing the software development in the early stage of SDLC. From 1986, the International Function Point User Group (IFPUG) has adopted and continued to enhance the Albercht Functional Size Measurements. FPA measures the functionality of software from the user requirements received or requested by the customer. FPA is the most efficient estimated software size metric than LOC [3]. FPA helps a lot in developing the product with utmost quality and increases the productivity. It is key driver to predict cost, schedule and other resources to build the software systems. FPA can be used for estimating maintenance effort in addition to estimating development effort. Functional Points (FP) requires a more detail examination of the user requirements or business requirements for effective estimation of cost or resources required to develop the software project. Functional Points can also be used in creation of best Test Suite for software project. Estimation based on FPA is usually done by specialized teams in organizations. Functional Points are useful for measuring the overall product deliverables at the beginning and at the end of a project. At beginning of the SDLC the function point can be used to estimate the overall schedule and cost of the project and at end it helps to measure the performance of the software. The representation of Functional Points in Fig. 8.1.

The following are the advantages of FPA

- It is purely language independent
- Counting of Functional Points derived from the first step (Requirement Step) of SDLC
- It provides reliable relationship to effort
- It can be more usable to GUI systems.

However, there are few disadvantages too. They are

- Most of the industries still use LOC as a size metric and not FP.
- It is purely time consumed approach.
- For research, scarcity of standard datasets.
- It is performed only after the design specification is done.

Fig. 8.1 Function points

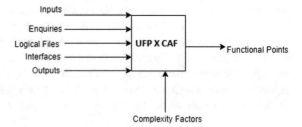

The major components of original Albercht's Function Points Analysis are

- Inputs (I): The number of external inputs that can be taken from outside the application boundaries. Eg: Product Code, Supplier Number, Date required etc.
- Outputs (O): The number of external outputs that can be sent outside the application boundary.
- Inquiries (Q): The additional information required along with the user inputs.
- Logical Files (L): The number of logical external files that includes various databases.
- Interfaces (F): The collection of data from the external systems.
- Complexity Adjustment Factors: These are the value-added external factors that affect the effort.

The Functional Points of a software project can be calculated using the following steps [6]:

1. Compute the scale for each Functional Points and scale varies from for 0 to 5. Where 0 is No Influence and 5 is Essential

$$F = scale * 14 \tag{8.1}$$

2. Determine the Complexity Adjustment Factor (CAF)

$$CAF = 0.65 + (0.01 * F) \tag{8.2}$$

3. Calculate Unadjusted Functional Point (UFP)
   Multiply each and every individual FP with its associated weighting Factor (WF) value as given in Table 8.1.

$$UFP = \sum I * WF + \sum O * WF + \sum Q * WF + \sum L * WF + \sum *WF \tag{8.3}$$

4. Calculate the adjusted Function Point (FP)

$$FP = UFP * CAF \tag{8.4}$$

Table 8.1 Weighting factors

Function type	Simple	Average	Complex
Inputs	3	4	6
Outputs	4	5	7
Inquiries	3	4	6
Logical files	7	10	15
Interfaces	5	7	10

8.2.3 Use Case Points (UCP)

Object Oriented Modelling is now-a-days very popular in the software industries for designing the software projects effectively. In the Unified Modelling Language (UML), both static and dynamic diagrams are available. The use case, sequence and component diagrams represent the functional requirements of the software projects. Effort required to design, and development of software projects are estimated in the early stage of SDLC. Software size and complexity factors are the drivers for predicting the software effort. One of the software size metrics in the object-oriented paradigm is the Use Case Points (UCP) [7, 8] which are derived from Use Case diagrams. UCP was firstly described by Gustav Karner in 1993 as part of his Diploma Thesis. UCP approach is an extension of Albercht FPA and other models which are derived from FPA. The number of Use Case Points in the software project consists of

- The number of Use Cases (UCs) with its associated complexity weights
- The number of Actors with its associated complexity weights
- Various Non Functional requirements
- Environmental Factors

UCP calculation Steps [9]

1. Calculate the Unadjusted Use Case Weight (UUCW): UCs are classified into Simple, Average and Complex. Each category is assigned a weight with 'Simple' having the lowest weight and 'Complex' having the highest weight. These weights are then multiplied with the number of actors in each category. Finally, these products are summed to derive the UUCW.
2. Calculate the Unadjusted Actor Weight (UAW): Actors are classified into Simple, Average and Complex. Each category is assigned a weight with 'Simple' having the lowest weight and 'Complex' having the highest weight. These weights are then multiplied with the number of actors in each category. Finally, these products are summed to derive the unadjusted actor weights.
3. Calculate Unadjusted Use Case Points (UUCP):

$$UUCP = UUCW + UAW \qquad (8.5)$$

4. Find Technical Complexity Factors (TCF); TCF will be determined by thirteen technical factors and each of them have weights from 0 (low Impact) to 5 (High Impact). The formula is given

$$TCF = 0.6 + (0.01 * Tfactor) \qquad (8.6)$$

where Tfactor is the weighted assessments of thirteen technical factors summed up.

5. Adjusting For Environmental Factors (EF): Environmental factors can also impact the size of the software. There are 8 EFs and each and every factor has its weights. These weights are summed up to determine the EFactor. The formula for calculating EF is as follows

$$EF = 1.4 + (-0.03 * EFactor) \qquad (8.7)$$

6. Calculate Final Use Case Point (UCP):

$$UCP = UUCP * TCF * EF \qquad (8.8)$$

The Technical Factors (TF) and Environment Factors (EF) are given in Tables 8.2 and 8.3 respectively. Consider the project with 350 Use case points and compute the duration to complete the project. Karner proposed that the man-hour per each use case is 20 h. Using this, the project with 350 use cases is translated into 7000 h required to develop the project. Ribu in 2001 analyzed and reported that this effort can range from 15 to 30 h per each use case. Schneider and Winters in 1998 proposed a model which is purely depending on the 8 Environmental Factors (EF) and they determined that 20 to 28 h is required to develop the project. For this example, with total 350 use case points and considering Schneider and Winters model of 20 to 28 h per each use case, the effort is calculated as follows.

- Total of 350 Use Cases
- Total hours required is 7000–9800 h.
- A total of 8 developers include Coder, Tester, DBA, Architect, etc.

Table 8.2 Technical factors

Factor	Description	Weight
TF1	Distributed systems	2
TF2	Response time/performance objectives	1
TF3	End-user-efficiency	1
TF4	Internal processing complexity	1
TF5	Code reusability	1
TF6	Easy to install	0.5
TF7	Easy to use	0.5
TF8	Portability to other platforms	2
TF9	System maintenance	1
TF10	Concurrent/parallel processing	1
TF11	Security features	1
TF12	Access for third practices	1
TF13	End user training	1

Table 8.3 Environmental factors

Factor	Description	Weight
EF1	Familiarity with development process used	1.5
EF2	Application experience	0.5
EF3	Object-oriented experience of team	1
EF4	Lead analyst capability	0.5
EF5	Motivation of the team	1
EF6	Stability of requirements	2
EF7	Part-time staff	−1
EF8	Difficult programming language	−1

- Developer spend 25 h of their time on development so that 8*25 h = 200 h/week or 400 h per Iteration.
- Divide 7000 h or 9800 h by 400. Then total 18 to 25 Iterations are required to develop the project.
- Depending on the scope of the project, configuration, installation etc., one or two iterations extra may be needed for development (20–27 Iterations).

8.2.4 Story Points

Story Point [10] is a popular size metric in Agile methodologies for the prediction of the effort for each and every User Story in the Product Backlog list. Story Point is an ultimate representation of the effort to develop each user story and it is done purely by Agile Teams. When estimation is done by story points, we should consider the amount of task to be done, the complexity of work and Risk factors that influence the development effort. Each User Story should have its business value or complexity and measure its size in story points like 1, 2, 3, 5, 8, 13, 20, 40, etc. Most of the Agile Teams will measure the user stories in 3, 5, 8 and 13 only. In Agile environments software is delivered to customer in Iterations or in Sprints. Each Sprint is time boxed iteration of 2 weeks or 4 weeks long where it completes 20–28 high priority story points. In Scrum Model, before start of the sprint, the sprint planning is attended by the product owner, scrum master and scrum team as shown in Fig. 8.2. The product owner fixes the priority of each and every user story of the product backlog. Each sprint is like a mini-project. During the sprint no changes are made that would affect the sprint goal. The development team members are the same throughout the sprint. An example on cost and schedule estimation using Story Points is shown in Table 8.4.

Many of the software projects fail due to inaccurate estimations. There has been a lot of research happening towards software estimations to estimate the effort accurately. Agile estimations using user stories are giving better results. In scrum, the whole team as shown in Fig. 8.3 estimates overall size of the user story in terms

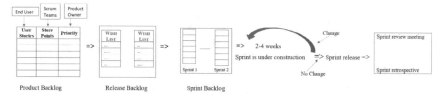

Fig. 8.2 Scrum model

Table 8.4 Estimation using story points

Estimated velocity	10
Total story points in project	80
Total sprints required	$80/10 = 8$
Uncertainty buffer	15%
Rework buffer	10%
Pre-release sprint	1
Total sprints required	11
Each sprints may take	4 weeks
Total duration of the project	$11 * 4 = 44$ weeks
Each sprints may take	Rs 15 lakhs
Total budget of the project	$11 * 15$ lakhs $=$ Rs 165 lakhs

Fig. 8.3 Scrum teams

of story points. Size can include effort, complexity and uncertainty. Scrum master facilitates the resources of what the scrum team requires. Product owner should be available to clarify the product backlog and as the right to prioritize the user stories. Planning poker technique is very popular and widely used in industry for their effort estimation. Planning poker uses playing cards with numbers on them as shown in Fig. 8.4. The numbers which are often based on Fibonacci series represents story points. A set of cards is given to each planning participant (Scrum Teams) and scrum master provides the details about user story.

Estimate high business relative size of product backlog items using planning poker (other techniques). Estimate your velocity using past details or by doing Sprint Planning Meeting. The estimation Include appropriate rework buffer (10%), to account for uncertainty buffer (15%) and pre-release sprint.

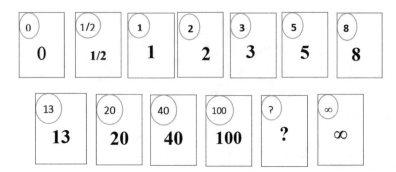

Fig. 8.4 Planning poker cards

8.3 Effort Estimation Techniques

Effort Estimation Process starts from gathering user requirements and building the scope of the project. The scope of the project leads to a business case which includes estimated time, budget, manpower and other supported environments required to complete the project successfully. Basically, there are three types of effort estimations which are very popular, Expert-Based Estimation, Algorithmic-Based Estimation and Soft Computing Techniques [11].

8.3.1 Expert-Based Estimation

The process of estimating the software development effort in the conventional days was commonly done by Expert Judgement Based [12] or Analogy Based Estimation. The Expert estimation technique is an informal approach as it does not imply any algorithm and is based on expert opinions, of one group or higher officials or from past experiences of peers. Expert Judgement will be made by various software professionals with different roles and responsibilities and with different project experience. Expert-Based estimation is purely derived from the scope of the projects. COCOMO or Functional Points is estimated based on mathematical formula whereas Expert Judgement understand the estimation process by the work break down structure which is purely based on what the expert believes as effort to complete total project. The Delphi technique, Activity based models and Work Breakdown Structure come under Expert estimation techniques. Expert Based estimation involves pure "gut feelings" to structure of project, past data or similar data about the project and checklist-based estimation process. In Agile Environments, estimation is done from Planning Poker is also based on Expert-Based estimation process [13] but here experts are scrum teams.

8.3.2   Algorithmic-Based Estimation

Effort, Duration and Cost of the software projects are estimated using the mathematical equations in Algorithmic-Based approach. Algorithmic Models are conventional models that were more popular [14] in the earlier days. These models include COCOMO [15], SLIM [16], SEER-SEM [17], ESTIMACS [18], Jensen Model [19], FPA [5], Soft Cost [20] etc. Generic effort estimation model like COCOMO which is presented by Barry Boehm uses the lines of code and/or functional points as key drivers for effort estimation. COCOMO Model provides clear and consistent results. The estimation using the above two models has both advantages and disadvantages. In some cases, it failed to predict the software effort accurately in situations where the data is ambiguous, uncertain, vague or ambiguous. Under these circumstances, soft computing techniques were considered good for SEE.

8.3.3   Soft Computing Techniques

The area of Soft Computing (SC) is delivering consistently precise estimates when compared with the Algorithmic Models or Expertise-Based Models. The application of SC for different Software Estimations started around 1990's. Soft Computing is a collection of different practices including Fuzzy Logic, Neural Networks, Adaptive Neuro Fuzzy Inference Systems, Regression Analysis, Support Vector Machines, Random Forest, Bayesian Networks etc. SC techniques resemble the biological processes more closely than traditional methods and are chiefly based on formal logical systems such as predicate logic. Using SC, the model learns from the training set of completed projects which then can be used for predicting with the other set of inputs. SC techniques are more suitable in all such situations, where there is a lack of impurity, uncertainty, ambiguity and clarity. These techniques have flexible information processing capabilities and delivers effective solutions at a lower cost.

8.3.3.1   Fuzzy Logic Approach

Initially, Fuzzy Logic (FL) [15], was the major emphasis for reliable experimentations. Fuzzy logic is a mathematical tool that is efficiently related to unnatural, uncertain and ambiguous data. Fuzzy rules, written in the form of if-then statements, map input space to the output space. For handling uncertain and unpredictable problems, most of the algorithmic models have been fuzzified.

Advantages:

1. Predicts when the data is imprecise and vague.
2. Fuzzy model is developed using normal human understandable language.

Disadvantages:

1. Harder to have a mark of meaningfulness. All the work done, needs to be redefined as the dataset changes.
2. Fuzzy Models cannot be generalized because the output obtained depends on the rules stored in the knowledge base.
3. Requires an expert to frame the optimized set of rules.

8.3.3.2   Neural Networks

Neural Networks (NNs) [21] depend on the principle of learning from experiences. NN consists of three entities, the neurons, the interconnection structure and the learning algorithm. The significant capability of NN is its ability to respond to new inquiries given dynamically. This method is more suitable in uncertain environments. In the last two decades, different types of NNs have been developed for software effort estimation. Back-propagation, trained feed-forward networks etc. are the NNs that are used majorly for estimations. The networks created consists of an appropriate set of neurons. A sequence of inputs is used as training data to train the network and achieve the correct output, thereby minimizing the prediction error. After completion of training, and assignment of values to weights, new inputs are given to the network for prediction. NNs appear like 'black boxes' as there is no information about the derivation of output. Hence, NNs can successfully be utilized for Software Development Effort Estimation.

Advantages:

1. NNs can estimate any non-linear relationships and is therefore useful for problems where there is a complex connection between input and output.
2. To construct NNs, many algorithms are available.

Disadvantages:

1. A clear guidance on the process of designing NNs is unavailable.
2. Requires a larger training dataset for high accuracy.
3. They are considered a black box—once given some input; there is a need to accept the generated output.

8.3.3.3   Adaptive Neuro-Fuzzy Inference System (ANFIS)

The combination of FL with NNs yields the hybrid Neuro-Fuzzy Systems [22]. ANFIS developed by Jang is one of the hybrid Neuro-Fuzzy Inference Systems that works in Takagi-Sugeno model. However, the ANFIS resembles the multi-layer feed forward NN, in which links indicate the direction of the signal between the nodes and no weight is attached. The normal hybrid neuro fuzzy system depends on the input given to a NN which changes completely on the input layer to true values. For the prediction of effort, this is an innovative field that gives good results. The conception

behind the model indicates that neural nets receive meaningful and crisp inputs that improve the overall quality of the predictions and the output. Neuro-fuzzy systems are suitable for a more accurate estimate of software projects efforts.
Advantages:

1. More efficient compared to Neural Networks.
2. Unacceptable rules can be removed.
3. Creating this model is serene.

Disadvantages:

1. Comparatively, a very new area.
2. Slightly time consuming.

8.3.3.4   Support Vector Machines

Support Vector Machine (SVM) [23] is a supervised learning algorithm that looks at various kinds of data and sorts them into their specialized category. It is an effective method for predicting the unknown data. SVM is mainly used for classifying data into two different segments depending on the features of the data. SVM trains and studies label data and then classify any new input data depending what it learns in the training phase. The main advantage of SVM is applicable for both Classification and Regression problem. Mainly SVM is used for Classification problems and Sup-port Vector Regression (SVR) handles the Regression problems. It also handles Non-Linear data with the help of SVM Kernel Function. Kernel Functions transforms the one-dimensional data to two-dimensional data (1D to 2D). SVM separates the data by using hyperplane. In SVM, the distance between the points and lines should be as far as possible. In technical terms the distance between the support vector and the hyperplane should be as far as possible. The support vectors are the extreme points in the datasets and hyperplane has the maximum distance to the support vectors of any class.
Advantages:

1. Can be used for both classification as well as for regression.
2. SVM works well when the distance between the support vectors and hyperplane is very large.
3. Its works well when there is a clear separation of classifiers.
4. It reduces risk of mis-classification.
5. It reduces the risk of over fitting.

Disadvantages:

1. It is not applicable for large datasets which contains more features where it takes long time for training the dataset.
2. Selection of Kernel function for solving regression problems is not so easy.

8.3.3.5   Random Forest

The Random Forest (RF) [24] technique was first developed by Leo Breiman in the year 1999. It is built on the CART (Classification and Regression Tree Algorithm) decision tree, which is the core building block of Random Forest implementation using machine learning techniques. Using Random Forest Techniques, the accuracy of prediction is higher, and the time complexity of training is less. Random Forest reduces the risk of over fitting and runs on large set of datasets. Random Forest or Random Decision Forest is a process of constructing a multiple decision trees when the training phase is developed. Decision tree is a tree-based diagram used to establish a course of actions. Each branch of tree is labelled with possible decisions, occurrences or reactions. Random Forest is an ensemble classifier which is made of using multiple decision trees. The large dataset is divided into multiple subsets of the dataset. Each and every subset of dataset is been represented by an individual decision tree. It helps to build a new model which uses past experiences. It is also an effective and accurate predictive model capable of performing Regression and Classification. In the process of determine the accuracy of prediction Random Forest Technique selects the certain number of features from all the available features. For each node of decision trees, compute the best partition set of features using YES/NO decisions or any other alternative approach. After completing the construction of multiple decision tress based on various variables, select the best path using more probable Yes or other decision actions.

Advantages:

1. It is most efficient and accurate prediction model.
2. It handle both Classification and Regression problems.
3. It suits best for large datasets.
4. It does not require any prerequisite preparation.
5. The construction of multiple decision tress is done parallel.
6. It reduces the risk of overfitting.

Disadvantages:

1. It may not be an efficient algorithm to train the non-smooth inputs like images and text.
2. It generates lot of decision trees while training the large datasets can have long training time and it utilizes a more memory.
3. It can tend to over fitting and overlapping.
4. It is a weak interpretable.

8.4   Effort Estimation Influencing Factors

Accurate Estimations of software effort and cost is great challenge for both organizations and researchers from the last two decades. Software size is one of the major input drivers for estimating the effort and cost, but it is not sufficient to estimate it accurately. Some external factors also impact a lot to accurate predictions of software effort. Few factors that determine the software development productivity has to be considered for estimating the software development effort. The following are some of the common factors that are influencing the software project effort [25].

Project Management: The key role in the software development is the Project Management (PM). PM is the main discriminator for success or failure of the software projects. The objective of the software project management is to deliver the software end product on time, within budget and meet all the requirements mentioned by the customer. But it can't ensure the delivery of software on time and within budget as the PM team fail to predict effort, analyze various risks and other factors that influence the software projects. The organizations should always consider the experience and skilled people in the PM team.

The Effect of Uncertainty: Predictors are supposed to have very good knowledge on requirements of the system to be implemented before estimating the software development effort. Estimators should use tools and techniques to estimate the effort accurately, which is very difficult to understand and use. In the conventional days the implementation of the software projects is in sequential approach and the estimation of the development effort is also somewhat easy. They used either expert based or algorithmic model for their estimations and none of them have given best estimations. Now-a-days, as the size of software and complexity of software dramatically changed, accurate SEE is difficult to achieve.

Software Size: The key input driver for all the estimation models is the software size. In the conventional days software can be measured either in Lines of Code (LOC) or Functional Point Analysis (FPA) and each of these have their own pros and cons. Now a days, most of the companies consider Use Case Points or Story Points. It is very hard to describe which metric is useful or appropriate in which environment.

Experience of the Estimation Team: The experience and skill set of the estimation team will be the main discriminator for accurate SEE of a software project. The domain knowledge of the estimating team will also influence the process of estimation. Generally, process of estimation is done by senior people in the organization those who have lot of experience in developing and maintaining several projects but not by the people who develop it. As they have lot of experience in estimating the software effort for a long run, they may ignore few factors while estimating the effort. It leads to failure of the software projects. In the Agile Methodologies Estimations are done by themselves who develop the software with the help of Scrum master which gives very good results in agile environments.

Dependency on Tools and Techniques: Estimation of the effort requires not only mathematical models but also the tools and techniques to develop them. Tools are used to automate the artifacts from one form to another form and techniques help the process of transformation. Most of the organizations are purely depending on automated tools and sometimes they too are not providing accurate results. Not only tools and techniques are required but also experienced and skilled people require for the better prediction.

Frequent Change of Requirements: Changing of requirements are quite common factor that influence the software productivity and thus impact the estimation of software project. From the conventional days of software development, the major problem faced by the industries are continuous change in requirements. In the early stages of SD, no customer gives complete requirements and they feel they have the flexibility to change the requirements at any stage of SDLC. Requirements form the basis for estimating size of software and help to estimate the effort/cost of software project. If the requirements are changing frequently then estimations too change. So, the requirements must be freeze or baselined before estimations.

8.5   Experimental Design

8.5.1   Evaluation Criteria

Examining the accuracy of the effort estimations using different models depends upon the evaluation criteria used. The performance metrics that can be effectively utilized for selecting the best model are given below [26]. A model with greater VAF and Pred is preferred. On the other hand, a low MARE, VARE and BRE model is preferred.

1. Variance Accounted For (VAF)

$$VAF(\%) = \left(1 - \frac{var(E - A)}{var E}\right) * 100 \qquad (8.9)$$

2. Mean Absolute Relative Error (MARE)

$$MARE(\%) = \left(\frac{\sum(f(R_E))}{\sum f}\right) * 100 \qquad (8.10)$$

$$where \ Absolute \ Relative \ Error(R_E) = \frac{|E - A|}{|E|} \qquad (8.11)$$

3. Variance Absolute Relative Error (VARE)

$$VARE(\%) = \left(\frac{\sum f(R_E - mean R_E)^2}{\sum f}\right) * 100 \qquad (8.12)$$

4. Prediction (n)
   Prediction at level n is defined as the % of projects that have absolute relative error less than n.

5. Mean Balance Relative Error (MBRE)

$$MBRE = mean\left(\frac{|E - A|}{min(E, A)}\right) \qquad (8.13)$$

6. Mean Magnitude of Relative Error (MMRE)

$$MMRE(\%) = \frac{1}{N}\sum_{i-1}^{N} MRE_i * 100 \qquad (8.14)$$

$$where\ MRE = \frac{|A - E|}{|A|} \qquad (8.15)$$

where N = No. of Projects, E = Estimated Effort, A = Actual Effort

8.5.2 Datasets

For the purpose of effort estimation, NASA 93 and Desharnais datasets are considered which are available publicly [27]. NASA 93 includes the 93 projects data taken from different NASA centers. Each project has 17 input parameters: the parameters are of of Intermediate COCOMO [26]. The attributes of NASA 93 dataset are Mode, size, required software reliability, data base size, product complexity, execution time, main storage constraint, virtual machine volatility, computer turnaround time, analyst capability, applications experience, programmer capability, virtual machine experience, language experience, modern programming, use of software tools, required development schedule. The dependent variable is the effort and all the above 17 input attributes are independent variables. The range of Mode is from 1.05 to 1.2. The range of size is from 0.9 to 980 KDSI. The range of dependent variable effort is from 8 to 8211 person-months.

The Desharnais dataset [27] consists of 81 instances with 12 attributes. The attributes are Project ID, Team Experience, Manager Experience, Year End, Length, Transactions, Entities, Points Adjust, Envergure, pointsNonAdjust, Language and Effort. The 10 attributes except Project ID and Effort are the independent variables that are used as input to the proposed Soft Computing Models. The dependent variable is effort. Effort is measured in person-months. The dataset is proposed by J. M. Desharnais in his Master's thesis in 1989.

8.5.3   Model Design

A few of the different Soft Computing techniques available in the literature are presented in Sect. 8.3.3. All these models can be used for estimating the software development effort. In this section, a detailed description on the development and working of Fuzzy Model, Neural Networks and ANFIS Model using a case study on the NASA 93 dataset and Desharnais dataset is presented and compared. The development of models using other soft computing techniques is left over for further experimentation.

8.5.3.1   Fuzzy Model for NASA 93 Dataset

Fuzzy Models can be used for solving all kinds of problems which are excessively tough to be understood comprehensively. Fuzzy Logic (FL) is an easy and simple way of mapping an input space to an output space using a specific set of rules. It has been very adequately used for Software Development Effort Estimation from the last decade. The analytical structure of a Fuzzy Logic Controller (FLC) comprises of 4 basic modules as shown in Fig. 8.5.

Fuzzification Module: Fuzzification is the first and most important operation to be done in an FLC, and it changes the range of input and output of the FLC into their respective Universe of Discourse (UOD). The second operation is to split the input into the respective linguistic variable. The parameters of this fuzzy module depend on the structure of the used membership functions (MF) [28].

Inference Mechanism: This module plays the next important role in FLC. The membership values, obtained in the previous step, are combined to get the firing strength for each rule. Then depending on this firing strength, the consequent part

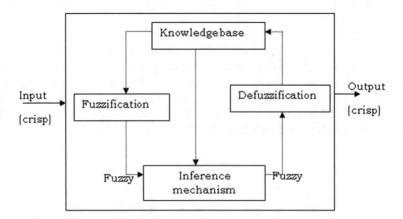

Fig. 8.5 Basic structure of FLC

Table 8.5 Results of various models for NASA 93 dataset

Model	VAF	MARE	VARE	Mean BRE	MMRE	Pred (25)%
Fuzzy model	99.5	3.5	27.1	0.11	18.66	89.4
RBNN	99.81	3.6	1.64	0.084	8.28	93.5
GRNN	98.48	18.6	11.24	0.339	16.83	82.6
ANFIS	99.9	4.68	1.8	0.1	10.16	94.1

of each fired rule is generated. The Takagi-Sugeno Fuzzy Inference System (TSFIS) is the commonly used interface mechanisms for solving engineering problems. The TSFIS is used for software effort prediction.

Knowledge base: It is considered as a database for the FLC. The key function of the knowledge base is to supply the required and necessary information for all the other modules.

De-fuzzification module: This module of FLC initially transforms the set of modified control output values into non-fuzzy values and then denormalizes the output that maps the range of values of fuzzy sets to the physical domain. For Sugeno type Fuzzy Inference System, the Defuzzification methods used are wtaver (weighted average) or wtsum (weighted sum).

The Fuzzy Model for NASA93 dataset was created in MATLAB using the genfis2 algorithm. The model was trained using 83 instances of the total dataset available. The model was tested using the entire dataset. A case study on the NASA 93 dataset presents the results of Fuzzy Model as per the evaluation criteria presented in Sect. 8.5.1. The experimentation results of the Fuzzy Model are shown in Table 8.5. The Fuzzy Model provided good results with VAF of 99.5% and Pred of 91.4%.

8.5.3.2 Neural Networks for NASA 93 Dataset

As NNs can be used to approximate any nonlinear function and depends primarily on learning from previous data. It can be utilized effectively for SEE applications as well. Artificial Neural Network (ANN) or Neural Network or Neural net in short [28, 29] is a large distributed parallel processor formed using different processing units called neurons which have the natural capacity for storing experimental knowledge and making it ready for use. NN corresponds to the brain in two aspects: First, through the process of learning similar to the brain, the knowledge is obtained by the network from the existing environment. Secondly, the acquired knowledge is stored in the form of an Interneuron connection strengths usually known as the synaptic weights. ANNs are successfully implemented for speech recognition, image analysis, controlling robots etc.

To create NNs, the number of neurons to be used and how these neurons are to be connected is to be finalized. Different types of NNs exists like the Feed forward NN, Back propagation NN, Radial Basis Neural Networks (RBNN), Generalized

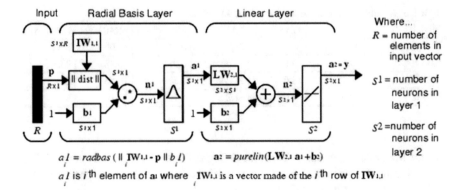

Fig. 8.6 Basic structure of RBNN

Regression Neural Networks (GRNN) etc. The literature [30, 31] shows that RBNN and GRNN provided better estimations for certain problems. Hence, in the present chapter, details of only these two models for SEE is presented. In general, the creation of any NN consists of six major steps:

- Identify the dataset to be used.
- From the entire dataset, pickup the training dataset (70–90%).
- Create the network.
- Train the network using the training dataset.
- Validate the network using the testing dataset.
- Use the network for similar other problems.

The Radial Basis NN comprises 2 layers: a hidden radial basis layer of S1 neurons, and an output linear layer of S2 neurons as shown in Fig. 8.6. This NN consists of Radial Basis Neurons and a Radial Basis Transfer function. An RBNN can easily be created in MATLAB using the available "newrbe" function. For the network created for the assessment of effort by using a case study on NASA 93 dataset, the input to the function is a matrix of order 17×83 and the output is a matrix of order 1×83. For this purpose, the spread value of 1.0 is used. The build RBNN has 83 neurons in the second layer along with the first layer. Simulation of the NN created is done using the 'sim' function. The input of the passed test data is the matrix of order 17 $\times$ 93 and the output is a 1×93 matrix.

The Generalized Regression Neural Network consists of 2 layers: radial basis layer and a special linear layer as shown in Fig. 8.7. GRNNs are frequently used for function approximations. It is very similar to RBNN with minute variation in the second layer. A GRNN can be easily created in MATLAB using "newgrnn" function available. For SEE using a case study on NASA 93 dataset using GRNN, the input and output to the "newgrnn" function are the same as that of "newrbe" function. For better results a spread of 0.49 is used.

For the present chapter, RBNN and GRNN for Software Development Effort Estimation are created using NASA 93 dataset initially. The obtained results are

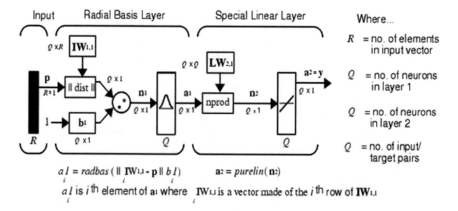

Fig. 8.7 Basic structure of GRNN

compared as per the evaluation criteria presented in the Sect. 8.5.1 as shown in the Table 8.5. The RBNNs provided better results with VAF of 99.81% and Pred of 93.5%. The GRNNs produced a VAF of 98.48% and Pred of 85%.

8.5.3.3 ANFIS for NASA 93 Dataset

The Fuzzy model and Neural Network limitations have been a key driving force behind the invention of intelligent hybrid systems, called the Adaptive Neuro Fuzzy Inference System (ANFIS) [32, 33]. The ANFIS used for software effort estimation is the Takagi-Sugeno Neuro Fuzzy Inference System (TSNFIS). It uses a blend of back propagation algorithm for learning the membership functions and least mean square method for determining the coefficients in the rule's consequents. The TSNFIS consists of six layers as shown in Fig. 8.8.

Layer-1 (Input Layer): In this layer, every node represents an input variable and just passes on the input values to the Fuzzification Layer.

Layer-2 (Fuzzification Layer): Each node in this layer corresponds to a member-ship function (MF) of the input variable. Output is the membership value that indicates the degree to which the input value is related to a fuzzy set. For each input variable, the number, type and size of the MF can be properly tuned by the clustering algorithm during the process of learning.

Layer-3 (Rule Antecedent Layer): The antecedent part of a rule, identified as a node represents the firing strength of the related fuzzy rule.

Layer 4 (Rule Strength Normalization): The ratio of a particular rules firing strength to the sum of all rules firing strength is calculated for every node in this layer.

Layer-5 (Rule Consequent Layer): Each node here is accompanied by a node func-tion that includes Layer 4's output and parameter set.

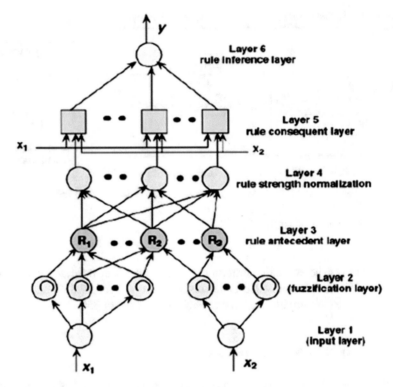

Fig. 8.8 Takagi-Sugeno neuro fuzzy system

Layer-6 (Rule Inference Layer): The final output of TSNFIS is the sum of all the incoming signals calculated by a single node in this layer.

As TSNFIS [24], has lowest Root Mean Square Error (RMSE) compared to the other Neuro Fuzzy models, it is used for Software Development Effort Estimation. For developing the TSNFIS Model in MATLAB, the "genfis2" function with a radius of 1.5 is used. Further to improve the accuracy, this FIS is trained using the "anfis" function available in the MATLAB toolbox. The model was trained using 83 records of the total dataset available and tested using the entire dataset. The experimentation results of a case study on the NASA 93 dataset compares the results of ANFIS Model with that of other models using the evaluation criteria presented in Sect. 8.5.1 as shown in Table 8.5. The ANFIS Model provided better results with VAF of 99.9% and Pred of 94%.

8.5.3.4 Soft Computing Models for Desharnais Dataset

As the SC Models Models have shown better estimations when compared to others in the case study on NASA 93 dataset, these models are further experimented using

Table 8.6 Results of various models for desharnais dataset

Model	VAF	MARE	VARE	Mean BRE	MMRE	Pred(25)%
Fuzzy model	91.25	11.84	4.35	0.20	18.43	79.0
RBNN	74.76	20.24	14.31	0.43	37.52	75.3
GRNN	78.95	24.30	10.06	0.39	33.64	67.9
ANFIS	91.29	11.26	4.20	0.19	17.85	82.7

another dataset called the Desharnais dataset available publicly. Out of 81 instances, total 70 instances are used for training all these models and all the 81 instances are used for testing. For designing the Fuzzy Model using genfis2, a radius of 0.7, squashFactor of 0.3, acceptRatio of 0.1, rejectRatio of 0.2 and default value 0 for verbose is used. The values of radius, squashfactor, acceptratio, rejectratio are varied and the results are checked. For the above values, the model have provided good estimates.

Similarly, for designing the RBNN, the function newrbe with a spread value of 26.2 is used. For designing the GRNN, the function newgrnn with a spread value of 22 is used. For designing the ANFIS, genfis2 with a radius of 0.46, squashFactor of 0.51, acceptRatio of 0.02, rejectRatio of 0.3 and default value 0 for verbose are used. The generated FIS is further tuned using the anfis function with training epoch number of 10, error goal of 0.4 and step size increment of 0.1 are used. The output of all the SC Models is the estimated effort. This estimated effort is compared with the actual effort as per the evaluation criteria presented in Sect. 8.5.1. The experimental results are presented in Table 8.6 and it is observed that ANFIS Model provided better software effort estimations when compared to other models.

8.6   Conclusion

The major problem prevailing in software industry is its incapability to predict the software effort and cost accurately. The conventional estimation models like Expert based judgement and Algorithmic models are not very accurate to estimate the effort and cost of the software project. In the present day, research experts and software companies are moving towards adopting Soft Computing techniques for reliable and accurate estimations. In this chapter, few of the soft computing techniques like Fuzzy Model, Neural Networks, Adaptive Neuro Fuzzy Inference System, Support Vector Machines and Random Forest Models have been introduced. For the experimental study, publicly available NASA 93 and Desharnais datasets are considered and Fuzzy Model, Radial Basis Neural Networks, Generalized Regression Neural Networks model and ANFIS Model are developed. The estimations of these models are compared using different metrics like VAF, MMRE, MARE, Mean BRE, RRSE, RAE and Pred (25)%. Comparing Fuzzy Model, Neural Networks and ANFIS, it was observed that ANFIS have provided better software development effort estimates. It can be concluded that ANFIS Model can be used in Software industry to estimate the Software Development Effort accurately at a low cost. Basing on the other case studies, experimentation can further be carried out with the different soft computing techniques like SVM, Random Forest, Bayesian Networks etc.

References

1. Rajkumar, G., Alagarsamy, K.: The most common success factors in cost estimation-a review. Int. J. Comput. Technol. Appl. **4**(1), 58 (2013)
2. Islam, S., Rokonuzzaman, M.: Process centric business case analysis for easing software project management challenges. JSW **6**(1), 20–30 (2011)
3. Hughes, B., Cotterell, M.: Software Project Management. Tata McGraw-Hill Education, New York (1968)
4. Trendowicz, A., Jeffery, R.: Appendix A: measuring software size, software project effort estimation (2014)
5. Symons, C.R.: Function point analysis: difficulties and improvements. IEEE Trans. Softw. Eng. **14**(1), 2–11 (1988)
6. Lavazza, L., Garavaglia, C.: Using function points to measure and estimate real-time and embedded software: experiences and guidelines. In: 2009 3rd International Symposium on Empirical Software Engineering and Measurement, pp. 100–110. IEEE (2009)
7. Nassif, A.B., Capretz, L.F., Ho, D.: Enhancing use case points estimation method using soft computing techniques. J. Glob. Res. Comput. Sci. **1**, 12–21 (2010)
8. Enachescu, C., Radoiu, D.: Software cost estimation model based on neural networks. In: Proceedings of the International Conference on Knowledge Engineering, Principles and Techniques, pp. 206–210 (2009)
9. Nassif, C.L.F., Bou, A., Danny, H.: Estimating software effort based on use case point model using sugeno fuzzy inference system. In: 2011 IEEE 23rd International Conference on Tools with Artificial Intelligence, pp. 393–398 (2011)
10. Hamouda, A.E.D.: Using agile story points as an estimation technique in cmmi organizations. In: 2014 Agile Conference, pp. 16–23. IEEE (2014)
11. Mendes, E., Mosley, N., Watson, I.: A comparison of case-based reasoning approaches. In: Proceedings of the 11th International Conference on World Wide Web, pp. 272–280. ACM (2002)
12. Kaur, P., Singh, R.: A proposed framework for software effort estimation using the combinational approach of fuzzy logic and neural networks. Int. J. Hybrid Inf. Technol. **8**(10), 73–80 (2015)
13. Kaur, A., Kaur, K., Malhotra, R.: Soft computing approaches for prediction of software maintenance effort. Int. J. Comput. Appl. **1**(16), 69–75 (2010)
14. Engel, A., Last, M.: Modeling software testing costs and risks using fuzzy logic paradigm. J. Syst. Softw. **80**(6), 817–835 (2007)
15. Babuška, R.: Fuzzy Modeling for Control, vol. 12. Springer Science & Business Media (2012)
16. Boehm, B., Clark, B., Horowitz, E., Westland, C., Madachy, R., Selby, R.: Cost models for future software life cycle processes: Cocomo 2.0. Ann. Softw. Eng. **1**(1), 57–94 (1995)
17. Boehm, B., Abts, C., Chulani, S.: Software development cost estimation approaches-a survey. Ann. Softw. Eng. **10**(1–4), 177–205 (2000)
18. Robin, H.: Using estimacs e. Management and Computer Services, Valley Forge, Pa (1984)
19. Jensen, R.W., Putnam, L., Roetzheim, W.: Software estimating models: three viewpoints. Softw. Eng. Technol. **19**(2), 23–29 (2006)

20. Leung, H., Fan, Z.: Software cost estimation. Handbook of Software Engineering and Knowledge Engineering: Volume II: Emerging Technologies, pp. 307–324. World Scientific, Singapore (2002)
21. Idri, A., Khoshgoftaar, T.M., Abran, A.: Can neural networks be easily interpreted in software cost estimation? In: 2002 IEEE World Congress on Computational Intelligence. 2002 IEEE International Conference on Fuzzy Systems. FUZZ-IEEE'02. Proceedings (Cat. No. 02CH37291), vol. 2, pp. 1162–1167. IEEE (2002)
22. Hodgkinson, A., Garratt, P.: A neurofuzzy cost estimator. In: Proceedings of the 3rd Conference on Software Engineering and Applications, pp. 401–406 (1999)
23. Prabhakar, M.D.: Prediction of software effort using artificial neural network and support vector machine. Int. J. Adv. Res. Comput. Sci. Softw. Eng. 3(3) (2013)
24. Satapathy, S.M., Acharya, B.P., Rath, S.K.: Early stage software effort estimation using random forest technique based on use case points. IET Softw. 10(1), 10–17 (2016)
25. Sehra, S.K., Brar, Y.S., Kaur, N.: Predominant factors influencing software effort estimation. Int. J. Comput. Sci. Inf. Secur. 14(7), 107 (2016)
26. Pvgd, P.R., Snsvsc, R.: Fuzzy based approach for predicting software development effort. Int. J. Softw. Eng. 1(1), 1–11 (2010)
27. Shirabad, J.S., Menzies, T.: The PROMISE repository of software engineering databases (2005)
28. Ying, H.: General siso takagi-sugeno fuzzy systems with linear rule consequent are universal approximators. IEEE Trans. Fuzzy Syst. 6(4), 582–587 (1998)
29. Sharma, V., Verma, H.K.: Optimized fuzzy logic based framework for effort estimation in software development. Int. J. Comput. Sci. Issues 7(2), 30–38 (2010)
30. Reddy, P., Sudha, K., Sree, P.R., Ramesh, S.: Software effort estimation using radial basis and generalized regression neural networks. J. Comput. 2(5), 87–92 (2010)
31. Reddy, P. et al.: Prediction of software development effort using RBNN and GRNN. Int. J. Comput. Sci. Eng. Technol. 1(4) (2011)
32. Sree, R.P., Reddy, P.P., Sudha, K.: Hybrid neuro-fuzzy systems for software development effort estimation. Int. J. Comput. Sci. Eng. 4(12), 1924 (2012)
33. Jang, J.-S.: Anfis: adaptive-network-based fuzzy inference system. IEEE Trans. Syst., Man, Cybern. 23(3), 665–685 (1993)

Chapter 9
White Box Testing Using Genetic Algorithm—An Extensive Study

Deepti Bala Mishra, Arup Abhinna Acharya and Srikumar Acharya

Abstract In unit testing phase, the developers test each module of the Software Under Test (SUT) by going through the source code of the software. This process of testing source code is called as white box testing. In white box testing, test cases are generated through different coverage based and failt based testing techniques. Related research has found that, among all coverage based testing techniques, the path coverage based testing can detect about 65% of defects in a SUT. The test data generated through path testing can also be exercised for mutation analysis i.e. path coverage based test data can be used for achieving highest mutation score during mutation testing. This chapter briefly reviewed some of the related research work on different testing techniques used in white box testing, such as path testing, and mutation testing. Different Genetic Algorithm (GA) based techniques have been developed for automatic test data generation and optimization in white box testing.

Keywords Test data generation · Test data optimization · Genetic algorithm · Path testing · Mutation testing · Critical path · Mutation score

9.1 Introduction

Software development undergoes a series of phases, known as Software Development Life Cycle (SDLC) shown in Fig. 9.1 [1]. Among all phases, software testing is an important phase of SDLC, as it is used to measure the quality of software and the the quality can only be ensured through rigorous testing [2]. So, qualitative, robust

D. B. Mishra (✉)
Department of Computer Science, C.V. Raman College of Engineering, Bhubaneswar 752054, India
e-mail: dbm2980@gmail.com

A. A. Acharya
School of Computer Engineering, KIIT University, Bhubaneswar 751024, India

S. Acharya
School of Applied Sciences, KIIT University, Bhubaneswar 751024, India

© Springer Nature Switzerland AG 2020
J. Singh et al. (eds.), *A Journey Towards Bio-inspired Techniques in Software Engineering*,
Intelligent Systems Reference Library 185,
https://doi.org/10.1007/978-3-030-40928-9_9

Fig. 9.1 Basic stages of
SDLC

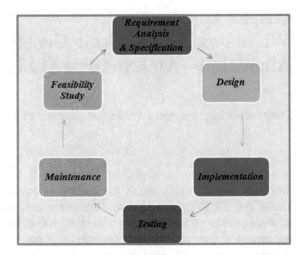

and trust worthy software can achieved an efficient testing methodology [3]. Testing
ensures whether the software meets the functional requirement, reliability, flexibility,
correctness etc. or not. Software testing is also defined as the process of exercising
or evaluating a system or a component of the whole system. It is also a process of
executing a software to find the errors present in it [3].

In software industry, there is a heavy loss about $500 billion per year due to
decrease in software quality and this is because of software failure [4]. Software
failures are caused by different faults and those faults can be identified in the software
testing. By successful testing the differences between expected and the actual results
can also be identified [5]. If the software does not perform as required and expected,
then a software failure is said to be occurred. A failure is an undesired behaviour
observed during the execution of the system being tested [6].

During software testing many issues like automatic generation of test cases, optimizing the test suite etc. are arises [7]. In order to address these issues, different
evolutionary techniques are used by researchers in last decades. Among all, Genetic
Algorithms (GAs) are frequently used by researchers to solve various issues found
in software testing fields, including automatic test data generation and optimization
[8].

The rest part of the chapter is organized in different sections as follows:

Section 9.2 describes some basic concepts, which are used in this chapter. In
Sects. 9.3 and 9.4, the basic definitions of black box testing and white box testing
are described respectively. A brief description on path testing is done in Sect. 9.5.
In Sect. 9.6 about mutation testing is discussed. Section 9.7 describes about how
different evolutionary algorithms and GAs are used in software testing fields. The
Section also describes some fundamental concepts on GA. In Sect. 9.8, a few related
research works on path testing using GAs as well as mutation testing using GAs are
presented. The conclusion of the chapter along with some future works are outlined
in Sect. 9.9.

9.2   Basic Concepts

This section presents some relevant concepts which are used in the rest of this chapter. For the sake of conciseness, we do not aim to present a detailed description of the background theory, instead, a brief introduction aimed at highlighting the basic concepts and definitions is provided.

9.2.1   Software Testing

Software testing is the process of resolving errors or faults present in a software product. It can also be defined as running the whole software or a portion of it with a given set of input [9].

Software testing is the process of producing correct, qualitative, and reliable software for the customer [1], but in our modern society, maintaining the quality of software is a very big issue. Due to decrease in software quality, many software industries are suffering with a heavy loss. Software failure is also play a vital role in decreasing the quality and these failures are caused by different faults [10]. For a complete validated and, high quality software, an efficient testing is required [9].

A Software Under Test (SUT) is tested in different levels including [11]: (i) unit testing which is done after a module or component of the software product has been coded and reviewed; (ii) integration testing which check whether the modules of the software, interface with each other properly or not; (iii) system testing where fully developed system is tested to assure that it fulfill all the requirements and specifications; (iv) acceptance testing which is done to confirm that the software meets its business requirements and specifications; (v) regression testing where an old test suite is to be executed along with the new test suite after each change to the system.

Basically, automatic test cases are generated by using two different testing techniques viz. *Static Analysis* and *Dynamic Analysis* [12]. A SUT is tested in two different ways as *Search Based Testing* and *Random Based Testing* [13]. Testing

- **Static Analysis**
  This method analyze the source code of the program and assess the properties without executing the SUT. The code inspection and code walk through are considered as static analysis method. Static analysis performs different tasks such as control flow analysis, data use analysis, interface analysis, and path analysis [14].
- **Dynamic Analysis**
  Dynamic analysis method executes the SUT with some input data and the actual result is recorded by the system. It collects information and evaluates test results through test execution. Test data are generated automatically with local search technique [15].
- **Random Based Software Testing (RBST)**
  In RBST, the program is tested with random inputs and check whether the expected

output is satisfied or not [16]. Randomly generated test data have high bug yield and the analysis of the test results is thus time-consuming [17]. However, this is a standard method through which other systematic methods can be compared to evaluate the efficiency [18].

– **Search Based Software Testing (SBST)**

In Search Based Software Testing (SBST) technique, the problem of searching an input datum is converted to an optimization problem. To deal with the optimization problem, many heuristic search algorithms such as Genetic Algorithm (GA), Particle Swarm Optimization (PSO), Ant Colony Optimization (ACO), Artificial Bee Colony Optimization (ABCO), and Simulated Annealing (SA) are widely used in software testing field [19]. This process starts with a generation of random test data and during the process the target is reached to a closer optimum solution and the above process is repeated until either the algorithm converges or the target solution is achieved [20].

9.3 Black Box Testing

In black box testing technique, test cases are designed using only functional specification based on input and output values of a program and it ignores the structural details of the software. There are various approaches to design test cases using black box testing like *Boundary Value Analysis (BVA), Equivalence Class Partitioning (ECP), Cause Effect Graphing (CEG), Pair Wise, State Based* [21].

9.4 White Box Testing

In white box testing, a unit or a module of the SUT is tested. It is also called as clear-box testing or glass-box testing. In structural testing, the developer test each module of the SUT, going through the source code of the software. During code testing, analysis of source code is required [11]. From code analysis the testing team can know how many number of test cases are needed to execute each and every statements of the SUT at least once. To test the source code in white box testing, test cases are designed in two different ways viz. *Coverage based testing* and *Fault based testing* [22].

9.4.1 Coverage Based Testing

In coverage based testing, different techniques are used to meet the test adequacy criteria. During test cases design process, the important factor is to select those test cases that matches a set of criteria [23]. Test cases are designed by taking various coverage factors of source code such as statement, branch, condition or decision, data flow, and path [24].

- **Statement Coverage Based Testing**
  It requires each and every statement of the SUT is to be executed at least once, but it does not guarantee to exercise the same statement in different flaws if any present. So this is a weak testing strategy [25].
- **Branch Coverage Based Testing**
  In this testing technique, test cases are designed to execute every decision at least once. One limitation of branch coverage based testing is that, it does not check for different sequences of the specific program [1].
- **Decision/Condition Coverage**
  This testing technique involves design of test cases to execute each component of a composite condition i.e. to execute all possible outcomes of a condition in a decision [26].
- **Data Flow Based Testing**
  This technique selects all possible test paths of a specific program, considering the definitions and uses of different variables present in program. This testing is basically used to test the programs containing nested if and loop statements [27].
- **Path Coverage Based Testing**
  In path coverage based testing each and every possible paths of the SUT are executed so that maximum errors can be detected [28]. The detail about path testing is discussed in Sect. 9.5.

9.4.2 Fault Based Testing

In fault based testing process, certain types of faults are detected. Mutation testing is one type of fault based testing, in which faults are inserted to make the original version of program mutated [3]. Mutation testing is a fault based testing. The details about mutation testing is discussed in Sect. 9.6.

9.5 Path Testing

In path coverage based testing each and every possible paths of the SUT are executed so that maximum errors can be detected. This testing technique is involved with the execution of all feasible paths in the program as there may be infinite numbers of paths due to presence of loops [16]. Path testing is the strongest among all as it satisfies all other type of coverage based testing techniques, by covering all feasible paths [6]. Path coverage based test cases can also be executed for 100% statement and branch coverage. So, it is the strongest coverage criteria among all coverage based testing techniques [25].

Path testing process consists of two major steps like target path generation and test data generation to cover the target paths. For target path generation, corresponding Control Flow Graph (CFG) is required that can be generated either manually or automatically using appropriate programming tool [7]. From CFG of a specific program, different logical paths can found. The test data which fulfill maximum path coverage criterion can be generated either randomly or in a heuristic approach. So, several algorithms are used for last decades to search for the required test data [16].

9.5.1   Control Flow Graph (CFG)

The CFG of a program represents the flow of sequence in a program through nodes and edges. The node is used for representing decisions and edges are used for control statements. From CFG, all linearly independent paths can be obtained [20]. An example is shown in Fig. 9.2, which represents the source code and CFG of finding the Greatest Common Divisor (GCD) of two inputted numbers.

9.5.2   Linearly Independent Path

Linearly independent paths are a set of paths, where no new path can be obtained through any linear operations of two existing paths in the basic path set. Linearly independent path for Fig. 9.2 are as follows: [3]

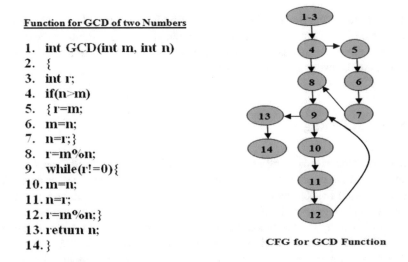

Function for GCD of two Numbers

1. int GCD(int m, int n)
2. {
3. int r;
4. if(n>m)
5. {r=m;
6. m=n;
7. n=r;}
8. r=m%n;
9. while(r!=0){
10. m=n;
11. n=r;
12. r=m%n;}
13. return n;
14. }

CFG for GCD Function

Fig. 9.2 Source code and CFG of the GCD program

- 1–3,4,5,6,7,8,9,10,11,12,13,14
- 1–3, 4,8,9,10,11,12,13,14
- 1–3, 4,8,9,13,14

9.5.3   Cyclomatic Complexity (CC)

McCabe [1] defined Cyclomatic Complexity (CC) as an upper bound to the independent paths present in a SUT. It can be found from the corresponding CFG, using using Eq. (9.1) [10]. So the CC for Fig. 9.2 is 3.

$$V(G) = E - N + 2 \tag{9.1}$$

where V(G) is the maximum number of independent paths in the CFG, E is total number of edges and N is total number of nodes present in CFG.

9.6   Mutation Testing

Offutt and Untch [10] developed mutation testing, a fault based white box testing. It evaluates and improves the quality of a test data by killing the active mutants present in the SUT [29]. In mutation testing process, the SUT is first tested with the test cases designed by any of the coverage based testing, and after the coverage based testing is complete, mutation testing is carried out. In this technique, mutation operators are inserted to make the program mutated [30]. The mutated program is tested by the test cases which are previously designed, to kill the mutants. If any of the test cases detect the mutant, then the mutant is said to be killed otherwise it is alive. Thus, the main objective of mutation testing is to select efficient test data having more error detection capacity, and to differentiate the initial program from the mutants [3].

Mutation testing is done through different steps as: [1, 31]

1. Construct the mutants for the Program Under Test (PUT)
2. Exercise the test cases with the mutation system to check the output of the program.
3. If the output is found incorrect, then a fault has been detected by a test case and the program must be modified to restart the mutation testing process.
4. If the output is correct, then the test case is executed against each live mutants.
5. If the output of a mutated program differs from the original program for the same test case, the mutant is assumed to be killed.
6. After each test case has been executed against each live mutant, the remaining mutants are either equivalent or killed. An equivalent mutant always produce the same output as the original program produce so no test case can kill it. But, in case of killable mutants, new test cases need to be created and the process iterates until all mutants are killed.

7. The efficiency of mutation testing can be evaluated in terms of higher Mutation Score (MS), which is defined in Eq. (9.2) [1].

$$MS = \frac{Total\ mutant\ killed}{Total\ mutants\ present\ in\ the\ program} * 100 \qquad (9.2)$$

9.7 Evolutionary Algorithms for Software Testing

Evolutionary Algorithms (EAs) are based on the principle of survival of the fittest and includes the modern heuristic search [32].

EAs are frequently used in software testing fields for automatic test data generation and optimization. The implementation of EAs achieves a great success on automatic test data generation for different coverage base testing. In this process the test adequacy criterion is formulated to an optimization problem, which is called as fitness function [5]. A set of individual solutions are generated and the best solution with highest fitness can be obtained through successive iterations [33].

Different EAs are developed including SA [5], GA [34, 35], Memetic Algorithms [5], PSO [36], ACO [37], ABC [38] etc. Among all EAs, GA have been successfully applied to a wide number of software engineering activities such as project planning, project cost estimation, automatic software testing, automated maintenance, and quality assessment etc. by providing valuable results in reasonable time [5]. In the context of software testing, the basic idea is to search the domain for input data which satisfy the goal of testing criteria. The testing problem is transformed into an optimization task, in which a numeric representation of the testing problem is required and a suitable fitness for the evolution of the generated test data can be derived [9]. For years, many researchers have proposed different GA based methods to generate test data automatically. It gives optimal test data using different operators through different generations in a complete automatic manner [32].

9.7.1 Genetic Algorithm (GA) for Software Testing

Genetic Algorithm (GA) is an optimization technique and based on the concept of surviving the fittest [39]. GA starts with a set of random solution, called as the initial population. Successively it forms the new population or new off-springs from the old population, based on the fitness value of the population towards the problem [32]. The more the fitness value the more they will be suitable to reproduce [6]. The process of new off-spring generation starts by the selection of parents on the basis of a certain selection process. After the process of parent selection, the new off-spring is generated using two different operators, as crossover and mutation successively [40]. In software testing environment, the most common method for generating test data automatically is the random generation method. In random method, test data are

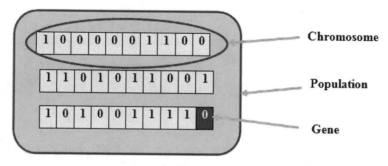

Fig. 9.3 Representation of chromosomes

generated uniformly without any knowledge of previous test data [41]. But, GA is an adaptive search method in which the test data is modified for achieving better fitness from one generation to the next generation i.e. it counts the knowledge of previous test data [42]. GA belongs to a class of directed search methods that are used for both modeling the system mathematically and solving optimization problems [43]. GA can effectively explore a sufficient portion of the solution space and gives an optimum result by setting the principal parameters in an effective way [44]. Through GA the best solution of the problem is searched from a population of solution candidates rather than a single. A set of individuals are generated randomly or heuristically, as the initial population. The individuals having better fitness are participating during selection and reproduction phase, and the others are likely to be discarded [45]. Optimal test data that satisfy the testing criterion can be generated through different generations and adjusted towards the optimum solution automatically [19].

GA can be defined by an 8-tupled expression which is given in Eq. (9.3) [7].

$$GA = (Co, F, Po, N, S, C, M, T) \tag{9.3}$$

In Eq. (9.3), Co = Coding type of individuals, F = Fitness value of individuals, Po = Population initially created, N = Range of population, C = Crossover type, M = Mutation type, T = Result of operation.

A. Representation of Chromosomes (Encoding): In GA, one possible solution to the problem is called individual, and all individuals together form a population. According to Goldberg [44], individuals are represented by strings, biologically called as chromosomes. Each chromosome contains a set of genes (bits in GA) which is shown in Fig. 9.3 [46]. Genes are represented in two different ways viz. binary coding and real coding.

– **Binary Encoding**

In this encoding method, parents are encoded by binary digits i.e. 0 and 1, which is shown in Fig. 9.4. Binary encoding have the capability of representing variety of information [32].

Fig. 9.4 Binary encoding

Fig. 9.5 Real encoding

- **Real Encoding**

   In Real encoding, the parents are purely real type, which is very simple and straight forward to use [46]. One example is shown in Fig. 9.5.

B. Selection Selection operator is used to select the best parents for performing other GA operations with a hope that, their offspring will give higher fitness in further generation [47]. Different types of Selection methods are used in GA, such as stochastic universal sampling, roulette wheel, linear rank, random, exponential rank, binary tournament and truncation etc. [6].

C. Crossover After selecting the better individuals by using selection operator, the crossover is applied to the chromosomes. It is the process of mating and reproducing new chromosomes, in which two individual swaps genes or sequence of bits between them on satisfying the probability factor of the operator [48]. Although, selection is surprisingly powerful, but it does not find new individuals in the search space. To obtain new individuals crossover and mutation operations are used. The crossover roughly mimics biological recombination between two single chromosome organisms [19]. Different types of crossover operators are used in GA like single point, two point, uniform, average and arithmetic [48].

D. Mutation This process is used to maintain genetic diversity in the population by altering chromosomes to introduce new good traits. Mutation is per- formed after crossover, if the mutation probability is true for the given iteration [49]. Different types of mutation operators are used in GA such as Bit string or one point, flip bit, boundary, uniform, non-uniform, insertion and Gaussian etc. [32].

E. Fitness Function Fitness is the objective function, measured by decoding a chromosome into the corresponding variables. In GA, the individual's performance is evaluated by the fitness function [32]. The fitness function, also called as objective function provides the useful information of each and every individuals. Through fitness, the search process is guided to find the global optimum solution as soon as possible i.e. in less number of generation [19].

F. Elitism In elitism process, redundant chromosomes are removed and a small proportion of the fittest candidates are copying into the next generation [50].

G. Work flow of GA The work flow of GA is shown in Fig. 9.6.

9.8   Related Work

In literature, a lot of works has already been done in the different field of white box testing using both static analysis methods as well as dynamic analysis methods. Different EAs are frequently used to address the issues found in white box testing like path testing, and mutation testing. In this section, some of the relevant research from the domain of automatic test data generation and optimization for path testing using GAs, mutation analysis through path testing using GAs are presented.

9.8.1   Test Case Generation and Optimization for Path Testing Using GA

Lin et al. [47] developed a GA based approach to generate path coverage based test data automatically. Test cases are generated to traverse a selected path. Authors have used extended Hamming Distance (EHD), where n is greater than 1 to measure the path distance. The proposed fitness, SIMILARITY can calculate the distance between the actual path executed and the target path. The proposed approach shows that test data can generate automatically to cover the most critical path, but the approach requires a huge number of test data to obtain target paths.

Mansour et al. [4] presented two stochastic search algorithms like SA and GA for generating test cases to execute paths of a specific program and their experimental result shows that SA tends to perform slightly better than GA in terms of success rates and execution time. Authors have used a weighted hamming distance between the operands of each predicate. The success rate for the proposed method is only 65%.

Chen et al. [51] developed a Multi Population GA (MPGA) method which selects individuals for free migration based on the fitness values. They have taken one search based program as Triangle Classifier Problem (TCP) for experiment and found MPGA based method can generate efficient test data for path testing than simple GA based method. They have designed their fitness by taking the branch distance between two paths. The proposed approach can achieve full path with large number of test data generation count.

Hermadi et al. [39] designed a GA based test data generator for multiple path coverage. The proposed technique can generate multiple test data to cover target paths in one GA run. Authors have extended their work [13] by calculating the sum and normalize the intermediate fitness value. The fitness is designed considering different attributes such as building blocks, normalization, neighborhood influence, path traversal method with rewarding, and weight adjustment. Their proposed technique has been compared through number of variations of weight and reward values. The experimental results ensures that the proposed test data generator is more effective and efficient than other existing test data generators in terms of covering multiple target paths with less number of test data generation.

Fig. 9.6 Work flow of GA

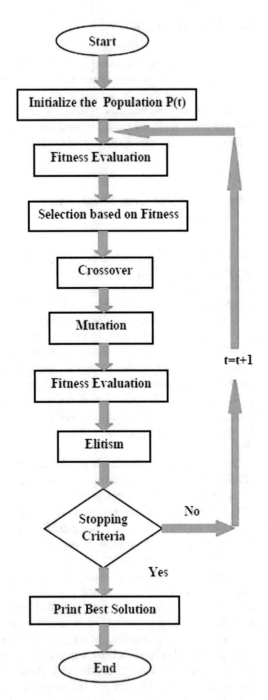

Srivastava et al. [6] presented a GA based approach for generating path coverage based test data automatically. The proposed approach can identified the most difficult path of a program under test. They have applied a 80–20 rule and the value 80 is given to the edge containing loops and conditions and 20 is given for each sequential edge. The path criticality can be found by calculating the total weight value. The criticality of a specific path depends on predicate, loop and branches present in the CFG.

Mala et al. [18] proposed a hybrid GA based method for test case optimization, by combining the features of GA and local search technique. The proposed approach is applied for test case generation and optimization of both path coverage and mutation analysis. Authors have taken eighteen numbers of academic and industrial programs written in C++/Java, such as binary search tree, coffee vending machine, stack operation, queue operations, library management system etc. to perform the experiments. From the reported result, it is seen that the proposed HGA based method produces both local and global optimal solutions with rapid convergence in comparison to GA and bacteriologic algorithm. It is also reported that the number of test cases is reduced to achieve maximum path coverage as well as mutation score for a SUT.

Shimin et al. [7] proposed a GA based method to generate path oriented test data automatically and reduces the redundant data to guarantee the testing efficiency. Authors have taken the most search based problem as TCP with MATLAB environment for their experiments. The designed fitness is adjusted the test data number dynamically, so that the path with narrow input fields can be generated, but the space occupied is very large when more parameters are used. From the experimental results it is seen that the run count is very high for generating the test data to cover all paths.

Maragathavalli et al. [52] proposed a GA based approach for multiple paths coverage by using dynamic instrumentation, to analyze the source code of a SUT. The proposed method is named as Path Reuse Method (PRM), which reduces the number of generation and execution time in GA process. The proposed method is implemented in the Java based program after instrumented in AspectJ tool. The test inputs are automatically generated to test the aspectual behavior of object oriented programs. Authors have taken seven different Java based programs like bubble sort, TCP, quick sort, BFS, queue, D-linked list, and report generation. The reported results shows that the proposed approach can generate test cases for multiple paths with less number of generation, which also supports to less time consumption.

Ghiduk et al. [53] proposed a new technique based on variable length GA to generate a set of basic test path automatically, in which the length of each chromosome varies from iteration to iteration. The set of basic paths can be used in any path testing method. From the experimental results, it is observed that the proposed approach could not achieve 100% path coverage.

Zhang et al. [12] proposed a GA based method to generate test data for covering multiple paths at a time. The proposed method can also detect the fault present in the SUT. A weighted GA based model is constructed to solve their multi-objective optimization problem. The proposed method is applied with several real world programs. They have taken bubble sort program and Siemens suite programs like print tokens, replace, schedule, tcas etc. which are written in C language. The proposed technique is compared with different methods like random method, Ahmed method [39], and Gong method [54], and the reported results confirms that the proposed technique can generate test data for traversing target paths as well as detecting maximum faults lying in the SUT.

Yao et al. [42] developed a hybrid system for multi-path coverage. The proposed approach combines the features of GA and local search method. Authors have designed the mathematical model for all target paths. But, the individual path is traversed by local evolution method. Authors have compared the proposed approach with Ahmed's method [39] and random method and the reported results ensures that the computational cost can be reduced by providing different granularity in different phases of fitness evaluation.

Zhu et al. [20] presented a GA based approach for generating multiple path coverage test data in SBST. An improved grouping strategy is proposed to balance the load of resources for traversing the target paths. Authors have used symbolic execution technique to reduce the search space of test data so that, the convergence process can be accelerated. The proposed method has been compared with Gong's method [54] and the performance result showed that, the efficiency is improved in terms of load balancing capacity.

Mishra et al. [40] presented a real coded GA based method to generate path coverage based test data automatically. The proposed method can also generate the test data to cover most critical paths present in a SUT. The proposed Real Coded GA can give an optimized test suite, which covers multiple paths at a time. The proposed method is named as Real Coded GA for Path Coverage (RCGAPC). The reported result confirms that, maximum path coverage can be achieved with less number of test data generation count.

Table 9.1 presents a summary of Literature Studies for software path coverage based test data generation and optimization using different EAs.

Abbreviations used in the summary table: (Pop-Initial population, Gen-Iteration, Sm-Selection, Cr-Crossover rate, Cm-Crossover type, Mr-Mutation rate, Mm-Mutation type, Em-Individual encoding).

9.8.2   Test Data Generation for Mutation Testing Using GA

Mala et al. [18] proposed a hybrid GA based method for test case optimization, by combining the features of GA and local search technique. The proposed approach is applied for test case generation and optimization both path coverage and mutation analysis. Authors have taken eighteen numbers of academic and industrial programs written in C++/Java, such as binary search tree, Coffee vending machine, stack operation, queue operations, library management system etc. to perform the experiments. Authors have used different mutation operators to calculate the mutation score. From the reported result, it is seen that the proposed HGA based method produces both

Table 9.1 Summary of literature studies carried on path testing using static and dynamic methods

Sl. No.	Authors	Method used	Parameters used	Results
1	Lin et al. [47]	GA	Gen-10, Pop-1000Em-Binary, Sm-Random, Cm-Two point, Cr-0.9, Mm-One point, Mr-0.02	GA can reduce the time required for lengthy testing of Software by automatic test case generation for path testing
2	Mansour et al. [4]	GA, SA, KA	En-Real encoding, Cr = 0.7, r = 0.1, Sm-Random	SA based method is able to test more paths than GA and KA.
3	Chen et al. [51]	MPGA (Multi Population GA)	Gen-400, Pop-1000En-Binary, Sm-Roulette wheel, Cm-one point, Cr-0.9, Mm-One point, Mr-0.02	MPGA can generate test data more effectively than Normal GA for path testing
4	Hermadi et al. [13, 39]	GA	Gen-100, Pop-30, En-Binary, Sm-Roulette Wheel, Cm-one point, Cr-0.5,0.9, Mr-0.1, 0.3	Results are more effective with multiple paths at a time
5	Srivastava et al. [6]	GA	Gen-3, Pop-4En-Binary, Sm-Random, Cm-Pair wise, Cr-0.8, Mm-flip, Mr-0.3	Testing efficiency becomes more by refining effort and cost estimation at the time of testing
6	Mala et al. [18]	HGA (GA+Local Search)	Gen-200, Pop-100, En-Binary, Cm-n-point, Cr-0.9, Mr-0.01	The proposed HGA based method produces both local and global optimal solutions with rapid convergence in comparison to GA and BA
7	Shimin et al. [7]	GA	En-Binary, Pop-100, Cm-One point, Mm-Uniform	The redundancy of test data is reduced and the efficiency of test data generation is improved
8	Maragathavalli et al. [52]	GA	Gen-2000, En-Binary, Pop-50,Sm-Random, Cm-Single point	The Path Reuse Method (PRM), can reduce the number of generation and execution time in GA process
9	Ghiduk, et al. [53]	Variable length GA	En-Binary, Sm-Roulette-wheel, Cm-One point, Cr-0.80, Mr-0.15	Proposed technique has the ability to generate feasible paths and causes a substantial reduction in path generation effort
10	Zhang et al. [12]	GA	–	The proposed technique can generate test data for traversing target paths as well as detecting maximum faults lying in the SUT

(continued)

Table 9.1 (continued)

Sl. No.	Authors	Method used	Parameters used	Results
11	Yao et al. [42]	GA	Gen-50,000, En-Binary, Pop-300 Sm- Roulette wheel, Cm-One point, Cr-0.6, Mm-One point, Mr-0.1	The proposed method needs less time consumption than random method
12	Zhu et al. [20]	GA	Gen-100, En-Binary, Pop-20, Sm-Roulette-wheel, Cm-One point, Cr-0.9 Mm-One point, Mr-0.3	The experimental result shows very good performance in generating a set of test data for multiple path in SBST
13	Mishra et al. [40]	GA	Gen-100, En-Real, Pop-50, Sm-Random, Cm-Average, Cr-0.9 Mm-Gaussian, Mr-0.3	Proposed RCGAPC shows very good performance in covering most critical path present in the SUT

local and global optimal solutions with rapid convergence in comparison to GA and bacteriologic algorithm.

Gong et al. [54] developed a GA based method to solve their formulated model and applied to several real world programs. The experimental result confirms that the proposed approach can generate test data both for traversing the target path and detecting faults lying in a SUT. Authors have inserted five different risk levels faults for doing mutation analysis such as Uninitialized Variable Fault (UVF), Out Of Bounds Array Access Fault (OBAF), Null Pointer Dereference Fault (NPDF), Illegal Computing Fault (ILCF) and Data Overflow Fault (DOF).

Zhang et al. [12] have presented a novel method to generate test data for covering multiple paths at a time. The proposed method can not only covers multiple paths at a time but also, detect faults present in the SUT. A weighted GA based model is constructed to solve their multi-objective optimization problem. The proposed method is applied with several real world programs like bubble sort and programs of Siemens suite as print tokens, replace, schedule, tcas etc. Authors have injected different faults like UVF, OBAF, NPDF, DOF etc. Their reported results confirms that the proposed technique can generate test data for traversing target paths as well as detecting maximum faults lying in the SUT.

Rani et al. [31] proposed a novel approach to generate test data and delete the active mutants using GA. Authors inserted different mutation operators like Statement Deletion (SDL), Constant Deletion (CDL), Operator Deletion (ODL), and Variable Deletion (VDL). An optimized test data is achieved by eliminating the redundant. The experimental result showed that, delete operator can saves huge time and effort during mutation analysis.

Khan et al. [55] proposed a GA based technique to generate test cases automatically. The generated test suite is further executed for mutation analysis in which mutation score is calculated according to the total number of mutants killed by each test suite. They have taken only one program as calculating the power value to perform the experiment. The experimental results shows that the proposed approach is able to find 100% mutation score with better test cases. Authors have proposed another method in [56], to optimize the efficiency of software testing by taking path coverage and boundary coverage based test data.

In [57], authors have proposed a Hybrid GA (HGA) based method to generate test cases automatically for mutation testing. They have taken the data flow information of a SUT for calculating mutation score. Authors have taken one program as generating prime numbers between two range for their experiments and a tool has been developed in C# to generate CFG for a program written in C Language. Author have compared their approach with both random method and GA based method and found that their proposed HGA based method produces maximum mutation score.

Bashir et al. [30] proposed a GA based approach which combines two way crossover with mutation method. The proposed method can generate better offsprings. Their improved GA is compared with random and normal GA method. They have implemented the approach with the tool as eMuJava V.2 with some selected mutation operators like Arithmetic Operator Replacement (AOR), Relational Operator Replacement (ROR), Absolute Value Insertion (ABS), Logical Connector Replacement (LOR), and Unary Operator Insertion (UOI). They de- signed the fitness by taking the different factors of the test case, like State Based Reach Ability Cost (SbRc), State Based Necessity Cost (SbNc) and Control Oriented Sufficiency Cost (CoSc). Authors have found that the proposed method can generate optimal test data with high mutation score which leads to reduce the computational cost.

Mishra et al. [41] proposed a GA based method for mutation testing. The proposed approach can generate test data for both path testing and mutation testing. Authors have designed their approach in a two phase manner. In first phase, 100% path coverage based test data are generated automatically. In the second phase, those test data are exercised for mutation analysis and an optimized test datum is found with highest mutation score. Fault Detection Matrix (FDM) is used to trace the faults for every test data. Table 9.2 represents the summary of Literature survey performed on the domain of Mutation testing using GA.

Table 9.2 represents the summary of Literature survey performed on the domain of Mutation testing using GA.

9.9   Conclusion

This chapter briefly reviewed some of the related research work on different testing techniques used in white box testing technique, such as path testing, mutation testing.

In path coverage based testing, test data are generated to cover the basic path of a specific SUT. The path coverage based test data can also be used to kill the mutants. It is observed that, different GA based techniques have been successfully applied for test case generation and optimization in software path testing. Researchers are also

Table 9.2 Summary of literature studies for mutation testing using GA

Sl. No.	Authors	Method used	Operators used to kill mutants	Results
1	Mala et al. [18]	GA	AOR, AOI, AOD, ROR, COR, COI, COD	The number of test cases is reduced to achieve maximum path coverage as well as mutation score for a SUT
2	Gong et al. [54]	GA	UVF, OBAF, NPDF, ILCF, DOF.	The method can generate test data to traverse the target path as well as detect the fault lying in the SUT
3	Zhang et al. [12]	GA	UVF, OBAF, NPDF, DOF	The proposed technique can generate test data for traversing target paths as well as detecting maximum faults lying in the SUT
4	Rani et al. [31]	GA	Delete operators like SDL, VDL, CDL, ODL	A prioritized set of mutants are generated which leads to reduce the cost and time of mutation testing
5	Khan et al. [55–57],	GA, HGA	–	The proposed technique generates better test cases to kill maximum number of inserted mutants
6	Bashir et al. [30]	GA	ABS, AOR, LCR, ROR, UOI	The proposed approach can find the optimal test cases in less number of attempts, which leads to reduce the computational cost and increase the mutation score
7	Mishra et al. [41]	GA	AOR, LCR, ROR	The proposed approach can find the optimal test data with highest mutation score in less number of test data generation

trying to achieve maximum path coverage through different GA based tools. The path coverage based test data can be used to achieve maximum mutation score using different GA based approaches. It is observed that, some of the methods can not only generate test data to cover a single or multiple paths, but also find the mutation score of each and every test cases for specific SUT.

From the existing related work, it is observed that the random based methods are not enough supported for robustness requirements. So, search based methods have been applied to address various issues found in white box testing process. The GA based test data generators have been applied for white box testing techniques viz. path testing, mutation testing and different fitness functions were also developed.

In future other EAs like PSO, ACO, SA can be hybridized with GAs for generating maximum path coverage based test suite and detect the criticality of a path during path coverage based testing as quickly as possible. Path coverage based test data can also be exercised for better mutation analysis.

References

1. Mathur, A.P.: Foundations of Software Testing, 2/e. Pearson Education, India (2013)
2. Sharma, A., Rishon, P., Aggarwal, A.: Software testing using genetic algorithms. Int. J. Comput. Sci. Eng. Surv.(IJCSES) 7(2), 21–33 (2016)
3. Mall, R.: Fundamentals of Software Engineering. PHI Learning Pvt. Ltd. (2018)
4. Mansour, N., Salame, M.: Data generation for path testing. Softw. Qual. J. 12(2), 121–136 (2004)
5. Malhotra, R., Khari, M.: Heuristic search-based approach for automated test data generation: a survey. Int. J. Bio-Inspired Comput. 5(1), 1–18 (2013)
6. Srivastava, P.R., Kim, T.-H: Application of genetic algorithm in software testing. Int. J. Softw. Eng. Its Appl. 3(4), 87–96 (2009)
7. Shimin, L., Zhangang, W.: Genetic algorithm and its application in the path-oriented test data automatic generation. Procedia Eng. 15, 1186–1190 (2011)
8. Mishra, D.B., Bilgaiyan, S., Mishra, R., Acharya, A.A., Mishra, S.: A review of random test case generation using genetic algorithm. Indian J. Sci. Technol. 10(30) (2017)
9. Ahmed, A.A.F., Shaheen, M., Kosba, E.: Software testing suite prioritization using multi-criteria fitness function. In: 2012 22nd International Conference on Computer Theory and Applications (ICCTA), pp. 160–166. IEEE (2012)
10. Chauhan, N.: Software Testing: Principles and Practices. Oxford University Press, Oxford (2010)
11. Nidhra, S., Dondeti, J.: Black box and white box testing techniques-a literature review. Int. J. Embed. Syst. Appl. (IJESA) 2(2), 29–50 (2012)
12. Zhang, Y., Gong, D.: Generating test data for both paths coverage and faults detection using genetic algorithms: multi-path case. Front. Comput. Sci. 8(5), 726–740 (2014)
13. Hermadi, I., Ahmed, M.A.: Genetic algorithm based test data generator. In: The 2003 Congress on Evolutionary Computation, CEC'03, vol. 1, pp. 85–91. IEEE (2003)
14. Khari, M., Kumar, P.: An extensive evaluation of search-based software testing: a review. Soft Comput., 1–14 (2017)
15. Huang, H., Liu, F., Yang, Z., Hao, Z.: Automated test case generation based on differential evolution with relationship matrix for IFOGSIM toolkit. IEEE Trans. Ind. Inform. 14(11), 5005–5016 (2018)
16. Mohi-Aldeen, S.M., Mohamad, R., Deris, S.: Application of negative selection algorithm (NSA) for test data generation of path testing. Appl. Soft Comput. 49, 1118–1128 (2016)
17. Biswas, S., Kaiser, M.S., Mamun, S.A.: Applying ant colony optimization in software testing to generate prioritized optimal path and test data. In: 2015 International Conference on Electrical Engineering and Information Communication Technology (ICEEICT), pp. 1–6. IEEE (2015)
18. Mala, D.J., Mohan, V.: Quality improvement and optimization of test cases: a hybrid genetic algorithm based approach. ACM SIGSOFT Softw. Eng. Notes 35(3), 1–14 (2010)
19. Mishra, D.B., Mishra, R., Das, K.N., Acharya, A.A.: A systematic review of software testing using evolutionary techniques. In: Proceedings of 6th International Conference on Soft Computing for Problem Solving, pp. 174–184. Springer (2017)

20. Zhu, Z., Xu, X., Jiao, L.: Improved evolutionary generation of test data for multiple paths in search-based software testing. In: 2017 IEEE Congress on Evolutionary Computation (CEC), pp. 612–620. IEEE (2017)
21. Shahbazi, A., Miller, J.: Black-box string test case generation through a multi-objective optimization. IEEE Trans. Softw. Eng. **42**(4), 361–378 (2016)
22. Desikan, S., Ramesh, G.: Software Testing: Principles and Practice. Pearson Education, India (2006)
23. Gong, D., Tian, T., Yao, X.: Grouping target paths for evolutionary generation of test data in parallel. J. Syst. Softw. **85**(11), 2531–2540 (2012)
24. Girgis, M.R., Ghiduk, A.S., Abd-Elkawy, E.H.: Automatic generation of data flow test paths using a genetic algorithm. Int. J. Comput. Appl. **89**(12), 29–36 (2014)
25. Mishra, D.B., Mishra, R., Acharya, A.A., Das, K.N.: Test case optimization and prioritization based on multi-objective genetic algorithm. Harmony Search and Nature Inspired Optimization Algorithms, pp. 371–381. Springer, Berlin (2019)
26. Godboley, S., Mohapatra, D.P., Das, A., Mall, R.: An improved distributed concolic testing approach. Softw.: Pract. Exp. **47**(2), 311–342 (2017)
27. Ghiduk, A.S., Harrold, M.J., Girgis, M.R.: Using genetic algorithms to aid test-data generation for data-flow coverage. In: 14th Asia-Pacific Software Engineering Conference, 2007, APSEC 2007. pp. 41–48. IEEE (2007)
28. Gong, D., Yao, X.: Automatic detection of infeasible paths in software testing. IET Softw. **4**(5), 361–370 (2010)
29. Silva, R.A., de Souza, S.D.R.S., de Souza, P.S.L.: A systematic review on search based mutation testing. Inf. Softw. Technol. **81**, 19–35 (2017)
30. Bashir, M.B., Nadeem, A.: Improved genetic algorithm to reduce mutation testing cost. IEEE Access **5**, 3657–3674 (2017)
31. Rani, S., Suri, B.: An approach for test data generation based on genetic algorithm and delete mutation operators. In: 2015 2nd International Conference on Advances in Computing and Communication Engineering (ICACCE), pp. 714–718. IEEE (2015)
32. Deb, K.: Optimization for Engineering Design: Algorithms and Examples. PHI Learning Pvt. Ltd. (2012)
33. Haga, H., Suehiro, A.: Automatic test case generation based on genetic algorithm and mutation analysis. In: 2012 IEEE International Conference on Control System, Computing and Engineering (ICCSCE), pp. 119–123. IEEE (2012)
34. Gupta, M., Gupta, G.: Effective test data generation using genetic algorithms. J. Eng., Comput. Appl. Sci. **1**(2), 17–21 (2012)
35. Kramer, O.: Genetic Algorithm Essentials, vol. 679. Springer, Berlin (2017)
36. Zhang, S., Zhang, Y., Zhou, H., He, Q.: Automatic path test data generation based on GA-PSO. In: 2010 IEEE International Conference on Intelligent Computing and Intelligent Systems (ICIS), vol. 1, pp. 142–146. IEEE (2010)
37. Mann, M.: Generating and prioritizing optimal paths using ant colony optimization. Comput. Ecol. Softw. **5**(1), 1 (2015)
38. Lam, S.S.B., Raju, M.L.H.P., Ch, S., Srivastav, P.R. et al.: Automated generation of independent paths and test suite optimization using artificial bee colony. Procedia Eng. **30**, 191–200 (2012)
39. Ahmed, M.A., Hermadi, I.: Ga-based multiple paths test data generator. Comput. Oper. Res. **35**(10), 3107–3124 (2008)
40. Mishra, D.B., Mishra, R., Das, K.N., Acharya, A.A.: Test case generation and optimization for critical path testing using genetic algorithm. Soft Computing for Problem Solving, pp. 67–80. Springer, Berlin (2019)
41. Mishra, D.B., Mishra, R., Acharya, A.A., Das, K.N.: Test data generation for mutation testing using genetic algorithm. Soft Computing for Problem Solving, pp. 857–867. Springer, Berlin (2019)
42. Yao, X., Gong, D., Wang, W.: Test data generation for multiple paths based on local evolution. Chin. J. Electron. **24**(1), 46–51 (2015)

43. Jena, T., Mohanty, J.R.: Disaster recovery services in intercloud using genetic algorithm load balancer. Int. J. Electr. Comput. Eng. **6**(4), 1828 (2016)
44. Goldberg, D.E.: Genetic Algorithms. Pearson Education, India (2006)
45. Garg, D., Garg, P.: Basis path testing using SGA & HGA with ExLB fitness function. Procedia Comput. Sci. **70**, 593–602 (2015)
46. Goldberg, D.E.: Real-coded genetic algorithms, virtual alphabets, and blocking. Complex Syst. **5**(2), 139–167 (1991)
47. Lin, J.-C., Yeh, P.-L.: Using genetic algorithms for test case generation in path testing. In: Proceedings of the 9th Asian Test Symposium, 2000, (ATS 2000), pp. 241–246. IEEE (2000)
48. Singh, G., Gupta, N., Khosravy, M.: New crossover operators for real coded genetic algorithm (RCGA). In: 2015 International Conference on Intelligent Informatics and Biomedical Sciences (ICIIBMS), pp. 135–140. IEEE (2015)
49. Umbarkar, A.J., Sheth, P.D.: Crossover operators in genetic algorithms: a review. ICTACT J. Soft Comput. **6**(1) (2015)
50. Das, K.N., Mishra, R.: Chemo-inspired genetic algorithm for function optimization. Appl. Math. Comput. **220**, 394–404 (2013)
51. Chen, Y., Zhong, Y.: Automatic path-oriented test data generation using a multi-population genetic algorithm. In: 4th International Conference on Natural Computation, 2008, ICNC'08, vol. 1, pp. 566–570. IEEE (2008)
52. Maragathavalli, P., Kanmani, S., Kirubakar, J.S., Sriraghavendrar, P., Prasad, A.S.: Automatic program instrumentation in generation of test data using genetic algorithm for multiple paths coverage. In: 2012 International Conference on Advances in Engineering, Science and Management (ICAESM), pp. 349–353. IEEE (2012)
53. Ghiduk, A.S.: Automatic generation of basis test paths using variable length genetic algorithm. Inf. Process. Lett. **114**(6), 304–316 (2014)
54. Gong, D., Zhang, Y.: Generating test data for both path coverage and fault detection using genetic algorithms. Front. Comput. Sci. **7**(6), 822–837 (2013)
55. Khan, R., Amjad, M.: Automatic test case generation for unit software testing using genetic algorithm and mutation analysis. In: 2015 IEEE UP Section Conference on Electrical Computer and Electronics (UPCON), pp. 1–5. IEEE (2015)
56. Khan, R., Amjad, M.: Optimize the software testing efficiency using genetic algorithm and mutation analysis. In: 2016 3rd International Conference on Computing for Sustainable Global Development (INDIACom), pp. 1174–1176. IEEE (2016)
57. Khan, R., Amjad, M., Srivastava, A.K.: Generation of automatic test cases with mutation analysis and hybrid genetic algorithm. In: 2017 3rd International Conference on Computational Intelligence & Communication Technology (CICT), pp. 1–4. IEEE (2017)

Chapter 10
Detection of Web Service Anti-patterns Using Machine Learning Framework

Sahithi Tummalapalli, Lov Kumar and N. L. Bhanu Murthy

Abstract Web services are being embraced by IT industry in the recent past to enable rapid development of distributed systems with optimal cost. Web services in SOA are self-adaptable to context, which makes SOA widely recognized in IT system as the technology, which has the potential of improving the receptiveness of both business and IT organizations. Web services help in building Service Based Systems (SBS) like Paytm, Amazon, Paypal, e-bay etc. which evolves frequently to fit the new user requirements which impacts the evolvability and quality of software design. Similar to software systems built using other paradigms, Service based systems also suffer from bad or poor design choices as in anti-pattern, code smells etc. **Anti-patterns** are explicit structures in the design that indicates violation of fundamental design principles and negatively impact the design quality. Anti-patterns have obstructive influence on the maintainability and perception of software systems. Thus there is a rising need for the early prediction of anti-patterns and refactoring them to improve the software quality in terms of execution cost, maintenance cost and memory consumption. In this work, a frame work is proposed for significant feature selection from source code metrics which includes Wilcoxon signed rank test, Univariate logistic regression analysis and Cross-correlation analysis. Then the different sets of features from various steps along with the original source code metrics are considered and are used for anti-pattern detection using 13 machine learning algorithms. Experimental results show the approximation capability of different classifiers and data balancing techniques with the features selected from the various steps of feature validation framework in addition to the original features for developing anti-pattern prediction model. The results also shows that the prediction model built with by the ensemble

S. Tummalapalli (✉) · L. Kumar · N. L. Bhanu Murthy
Department of Computer Science and Information Systems, Birla Institute of Science and Technology-Pilani, Hyderabad Campus, Jawahar Nagar, Hyderabad, Telangana, India
e-mail: p20170433@hyderabad.bits-pilani.ac.in

L. Kumar
e-mail: lovkumar@hyderabad.bits-pilani.ac.in

N. L. Bhanu Murthy
e-mail: bhanu@hyderabad.bits-pilani.ac.in

© Springer Nature Switzerland AG 2020
J. Singh et al. (eds.), *A Journey Towards Bio-inspired Techniques in Software Engineering*,
Intelligent Systems Reference Library 185,
https://doi.org/10.1007/978-3-030-40928-9_10

189

techniques using the features obtained from proposed feature selection framework outperforms other techniques.

Keywords Software engineering · Anti-pattern · Web service · Imbalanced data · Service oriented architecture · Machine learning · Prediction · Source code metric

10.1 Introduction

In software industry, it is important for the systems to be autonomous, heterogeneous and adaptable to the context, which lead to the evolution of web services based on service oriented architecture (SOA). SOA is the evolution of distributed computing for the successful business enterprises and modern public administrations towards the integrated working of specialist departments and information technology. Web services are used to implement SOA, so as to make the services accessible over the internet and this services are independent of underlying platforms and programming languages. These are self-adaptable to context which makes SOA widely recognized in IT systems as the technology, which has the potential of improving the receptiveness of both business and IT organizations. Web services can be deployed on the systems i.e., clients and servers with various platforms and this services are implemented in several languages. This web services are depicted by interfaces and can be accessed by using open protocols. Web services provides and implementers has to adhere with the usual standards to make the application, language and platform independent so that the offered services can be discovered and used by other applications. SOA service constitute of XML documents that are self describing and platform independent. WSDL helps the client program connected to a web service to determine the functions that are available on the server and acts a point of contact to the web service users.

The modeling of Service Based Systems (SBSs) like Paytm, eBay, Drop Box, Amazon etc., is made possible by the usage of SOA and the evolution of this systems leads to various issues. SBSs must progress to fit new user requirements, adapt new execution contexts such as addition of new devices and technologies. SBSs are prone to constantly change to indulge new user requirements and adapt the executional contexts, like any other large and complex systems. SBSs Quality of Service (QoS) and design may degrade the design because of all these changes and often result in a common poor solution to recurring problems, called Anti-patterns [1]. These are the structures in the design that indicates violation of fundamental design principles and reduces the quality of the design. Anti-patterns makes it hard for the evolution and maintenance of the software system but they are also known to help detecting problems in the architecture, code and the management of software projects. Studies revealed that the class having an anti-pattern is nine times more susceptible to fault proneness when compared to a class having no anti-pattern. Similarly, classes having anti-patterns have approximately three times higher probability of change proneness with respect to classes having no anti-Patterns. These days the study of detection of anti-patterns is an active research area which extends the study of design patterns into more extensive fields. Brown et al. [2] aiming the industrial and academic audience, provided a thorough study on anti-patterns, code smells and heuristics and he described a total of 40 anti-patterns in his book. In this experiment, following anti-patterns are considered:

1. Fine grained Web service (FGWS)
2. God object Web service (GOWS)
3. Chatty Web service (CWS)
4. Ambiguous Web service (AWS)
5. Data Web service (DWS)

In this work, a validation framework is proposed for feature selection, to identify a suitable set of metrics, out of a total 228 metrics, for anti-pattern detection. The prognostic performance of the preferred features or metrics is evaluated with the help of two ensemble techniques and eleven machine learning algorithms. The experiments are performed on a dataset downloaded from GitHub repository. From the conjectural results, it is concluded that the prediction model built using the features obtained from the proposed framework for the feature selection gave preferable outcome when compared to the model using all the original features. The results also shows that the Major Voting Ensemble (MVE) method gives better performance compared to all other classifier techniques. In this chapter, the proposed work addresses three different research questions:

- **RQ1: What is the capability of various data sampling techniques to predict web service anti-patterns?**
  In this work, to handle data imbalance problem, two different data sampling techniques have been considered. The considerability and dependability of theses techniques are computed on different anti-patterns using AUC and statistical test analysis.
- **RQ2: What is the capability of the features selected at various steps of proposed feature selection framework over original features to predict web service anti-patterns?**
  In this work, significant features, significant predictors and uncorrelated significant predictors selected using wilcoxon sign rank test, Uni-variate logistic regression and un-correlated feature analysis are considered. The significance and reliability are evaluated using AUC and statistical test analysis on different anti-patterns.
- **RQ3: What is the impact of various classifiers to predict web service anti-patterns?**
  In this work, 11 classifier techniques with 2 ensemble techniques have been considered to train the models for predicting web service anti-patterns. The considerability and dependability of these techniques are quantified on different anti-patterns using AUC and statistical test analysis.

10.2   Literature Survey

Over the last decade, research on anti-pattern detection has gained increased attention from software engineer researchers and practitioners. There are some studies reported for the web service anti-pattern detection but extensive work is done on the detection of anti-patterns in object oriented design but the techniques there cannot be implemented in the detection of anti-patterns in the web services because of the differences in their granularity level i.e., class versus system level. An analysis of the results of the various systems for the detection of anti-patterns is presented based on the varying techniques reported for the mining of software repository.

Detection and specification of anti-patterns in SOA and web services is a relatively new field. The problems of SOA anti-patterns is addressed only by a few works. Dudney et al. [3] wrote the first book on anti-patterns which provides the informal definitions about the set of web service anti-patterns. In later years, a range of SOA anti-patterns features were described. Furthermore, a variety of research work has been done, which addressed the issue of web service anti-pattern detection. Rodriguez et al. [4] performed automatic detection of common issues that should be steered clear of, while creating web services by presenting some heuristics. This heuristics are used for the discoverability and the removal of anti-patterns. Moha et al. [5] reckoning on a rule-based language at a higher level of abstraction to specify anti-patterns, has proposed an approach by name Service Oriented Detection of Anti-pattern (SODA) for anti-patter detection in Service Component Architecture (SCA) systems, but the approach proposed cannot deal with remote web services, it can only handle web services developed with plain java. Ali et al. [6] to detect anti-patterns has proposed an approach based on genetic programming. The proposed approach detects anti-patterns by producing some detection rules established from the metric and the threshold values.

Palma et al. propose a web-service anti-pattern detection algorithm that identifies the specific behaviors of the web service anti-patterns [7]. Their proposed approach was able to detect 10 anti-pattern and the approach was evaluated on the dataset from weather and financial domain. Nayrolles et al. found that at present there is a abundance of tools that detects anti-pattern automatically [8]. Hence, they proposed a prototype of a tool, i.e., SODA, that uses rule cards to specify anti-patterns. Some of the researchers used structural and behavioral data in UML designs, identified and predicted anti-patterns using quality metrics. An approach was proposed for determining all the occurrences of design patterns and anti-patterns in the graphical representation of the source code using a rule based matching approach. Some researchers examined the ways in which bayesian network can be used to improve upon the anti-pattern ontology in order to strengthen the already existing ontology-based detection process. Marinescu et al. [9] proposed detection strategies to detect and constraint the anti-patterns in systems. To detect 10 anti-patterns using this approach, 10 detection strategies are defined by the authors. The defined strategies has 2 pitfalls: (1) Without having deep knowledge of metric based rules, it is difficult for the user to successfully detect anti-patterns. (2) Different threshold values are leading to different results and the definition of the threshold is very difficult to understand. The approaches proposed by Marinescu et al. [9] depends mostly on the use of code or design quality metrics and the thresholds to detect the anti-patterns. Practitioners and software developers should have extensive knowledge about the code or design quality metrics and software quality, to use the above approaches to designate new anti-patterns or to re-engineer existing detection rules for the improvement of the performances. Derivation of the rules from the textual description/definition of anti-patterns requires thorough knowledge of the domain. The other issue of these approaches, is the availability of a selection of threshold values whose definition is difficult to comprehend.

10.3   Research Background

10.3.1   Web Service Definition Language (WSDL)

The resources in a SOA environment, are frequently offered as self-sustaining services that can be used by the web service users without the understanding of the implementation of the underlying platform. These services are issued in a manner that allows the developers to easily integrate them in the applications. Henceforth, the SOA relied systems are independent of the underlying platforms and developed technologies. When a web service is being implemented, some decisions like which transport, packaging protocol and network the service supports are to be made. All this decisions are represented using a description which helps the service consumer to contact and use that service. That description is called as Web Service Description Language (WSDL). **WSDL** is the standard format for describing the web service's interface and provides users with a point of contact. It is an XML based language used for transferring information in distributed environments and contains structured information about the web services.

10.3.2   Experimental Dataset

In this work, experiments are conducted on web-services that are available on the GitHub shared by Ouni et al. [6]. Because of the benefits of open source i.e., the obtainable data repositories and the source code openness, the dataset for the experiments is downloaded from tera-PROMISE repository. This gives the other researchers a chance to reproduce our work so that they can collate and gauge their approach with the proposed framework on the similar dataset. Dataset considered has 226 WSDL files, from which the source code metrics has to be extracted and the extraction process is discussed in the later section.

10.3.3 Source Code Metrics

Software metrics are statistical predictions and estimations, not just a number and are the tools for understanding varying aspects of the code base, and the project progress in software engineering. These metrics helps in acquiring duplicate objective measurements that can be handy for quality assurance, estimating costs, software performance, measuring productivity, assessing management and debugging. The role of the software metrics is to predict project success, project risks and to find defects in code prior to release and also after release. Some of the frequently used metric suites, indicating the quality of any software, are: Chidamber and Kemerer metrics, Abreu MOOD metrics suite, Henderson-Sellers, Bieman and Kang, McCabe complexity metrics etc. In this work, Chidamber and Kemerer metrics (CK metrics) which are widely used in research studies are computed, to measure the object oriented design of the WSDL file along with the quality of the design. The object oriented (oo) metrics at the web service level are computed using CKJM extended tool, which calculates the metrics by processing the byte code of the compiled java files. A total of 18 object oriented namely Depth of Inheritance (DIT), Weighted Methods per Class (WMC), Response for Class (RFC), Number of Children in Tree (NOC), Coupling Between Objects (CBO), Lack of Cohesion in Methods (LCOM), Efferent Coupling (Ce), Afferent Coupling (Ca), Number of Public Methods (NPM), Measure of Aggregation (MOA), LCOM3, Cohesion Among Methods of Class (CAM), Lines of Code (LOC), Data Access Metric (DAM), Inheritance Coupling (IC), Coupling Between Methods (CBM), Average Method Complexity (AMC) and Measure of Functional Abstraction (MFA) are calculated. The definitions of all these metrics along with there computational formulae are discussed in the CKJM documentation [10].

10.3.4 Aggregation Measures

Each WSDL file has more than one java class, for which the object oriented metrics at the class level are generated. But our focus here, is to detect anti-patterns at the file/system level. To do this, aggregation measures are to be applied on the class level oo metrics obtained from the java files to compute metrics at the system level.

Data points that represent a group average instead of information from an individual are called aggregates. Aggregates are produced by combining information from multiple sources. When you aggregate data, you use one or more summary statistics, such as a standard deviation, mean or median to provide a simple and quick description of some phenomenon of interest. Vasilescu et al. [11] claims that the aggregation of metrics from micro level (i.e., package, method or class) to macro level (i.e., program or file level) will help in pulling the outlier's into the substantial amount of data. Further, researchers determined that the results diverge based on the aggregation techniques that are being used and the correlation between them is not strong. Source code metrics are often computed at the method and class level but the

Aggregation Measure	Description	Computation Formula
Arithmetic Mean	The middle value of a feature	$\mu_x = \frac{1}{N}\sum_{i=1}^{N} x_i$
Median	Centre value of a feature	$M_x = \begin{cases} x_{N+1/2} & if\,N\,is\,odd \\ 1/2(x_{N/2} + x_{N+2/2}) & otherwise \end{cases}$
Standard Deviation	A measure of dispersion of a set of values from the mean	$\sigma_x = \sqrt{\frac{1}{N}\sum_{i=1}^{N}(x_i - \mu_i)^2}$
Variance	Shows difference of actual and expected value of a feature	$var(x) = \frac{\sigma_x}{\mu_x}$
Skewness	A measure of asymmetry in distribution	$\gamma_1 = \frac{\sum_{i=1}^{N}(x - \bar{x})^3/N}{(\sigma(x))^3}$
kurtosis	Measures the sharpness of the peak of a distribution curve	$\gamma_2 = \frac{\sum_{i=1}^{N}(x - \bar{x})^4/N}{(\sigma(x))^4}$
Gini Index	Statistical measure of distribution to measure the inequalities in the frequency distribution	$I_{Gini}(x) = \frac{2}{N\sum_x}[\sum_{i=1}^{N}(x_i * i) - (N+1)\sum_x]$
Hoover Index	A measure of income metrics	$I_{Hoover}(x) = \frac{1}{2}\sum_{i=1}^{N}|\frac{x_i}{\sum_x} - \frac{1}{N}|$
Theli Index	A measure of income inequality and other phenomena	$I_{Theli}(x) = \frac{1}{N}\sum_{i=1}^{N}(\frac{x_i}{\mu_m} * \ln(\frac{x_i}{\mu_m}))$
Atkinson Index	A measure of income metric inequality.	$I_{Atkinson}(x) = 1 - \frac{1}{\mu_x}(\frac{1}{N}\sum_{i=1}^{N}\sqrt{x_i})^2$
Generalized Entropy	A measure of income inequality in a feature set.	$GE_x = -\frac{1}{N\alpha(1-\alpha)}\sum_{i=1}^{N}[(\frac{x_i}{\mu_x})^\alpha - 1], \alpha = 0.5$
Shannon Entropy	An Average amount of information required to identify random sample from the distribution.	$E_x = -\frac{1}{N}\sum_{i=1}^{N}[\frac{freq(x_i)}{N} * \ln\frac{freq(x_i)}{N}]$

Fig. 10.1 Aggregation measures

project risks is often generated at the file-level [11]. Hence, aggregation measures are used in this work for determining the metrics at the file-level, which will help in anti-pattern detection at system level. Multiple aggregation measures are applied on the source code metrics computed, for risk prediction susceptibility and it is recommended, to use all aggregation measures to achieve better accuracy in predicting the anti-patterns. Therefore, a total of 16 aggregation measures are applied on each source code metric.

Aggregation measures are grouped into two categories: standard summary statistics and inequality indices. The first category presents quantitative data in manageable form whereas the second category is used to make inferences from our data to more generalized conditions. Standard summary statistics includes central tendency measures as *mean, median*; additive measures as *summation*; distribution shape measures as *skewness, kurtosis* [11]; statistical dispersion measures as *standard deviation, variance*. The second category includes econometric inequality indices, which are normally used to determine the dispersion of incomes, like Gini, Theli, Atkinson and Hoover indices. In this work, the summation measures are not considered but the proposed framework uses other measures like min, max, Quarantile1 (25%), Quarantile3 (75%) and entropy measures like Shannon entropy, Generalized entropy to measure the un-certainties associated with the metric values.

Let x_i be the value of metric **m** in the ith method in a file that has **N** methods and the computation formulae of each aggregation measure is shown in Fig. 10.1.

As discussed in literature survey, the research till now was being carried out using the static metrics, dynamic metrics, web service interface metrics, code metrics or a set of object oriented metrics at the class level which is different from the set of metrics level that is being considered in this work. However, to the best of our knowledge this is the first work, being carried out using a set of aggregation measures which are to be applied on each of the object oriented metric being considered for

SNo	Software Metric	P-value	SNo	Software Metric	P-value
1	WMC	0	10	LCOM3	0
2	DIT	0	11	LOC	0
3	NOC	2.71E-214	12	DAM	0
4	CBO	0	13	MOA	0
5	RFC	0	14	MFA	0
6	LCOM	0	15	CAM	0
7	Ca	0	16	IC	1.08E-206
8	Ce	0	17	CBM	1.08E-206
9	NPM	0	18	AMC	0

Fig. 10.2 Friedman test analysis

the prediction of web service anti-patterns. As per Fig. 10.1, the definition and the computation formulae used to measure each aggregation measure is different from the other. To verify whether all the aggregation measures being considered on each of the software metric are appreciably different from each other or not, the **Friedman Test** is conducted.

Friedman test is used to measure difference between groups when the dependent variable being measured is ordinal. The null-hypothesis for the Friedman test is that, the aggregation measures applied on the software metrics are not considerably unrelated, if the calculated probability i.e., P-value is below the selected threshold value which is 0.05. If the null-hypothesis is rejected, it can be concluded that aggregation measures for each of the software metric are considerably unrelated from each other. The results of Friedman test on the aggregation measures for each of the software metric are shown in Fig. 10.2. From the results it can been seen that the null-hypothesis is rejected. Hence its is concluded that, the aggregation measures used on each software metric are significantly unrelated.

Figure 10.3 shows the results for the wilcoxon test of aggregation measures for WMC and DIT software metrics respectively. Symbol '.' refers that there is no noteworthy correlation between the two metrics. Symbol '∗' signifies that there is a notable correlation between the two metrics and one of the metrics can be chosen to be ignored. From Fig. 10.3a, it can be inferred that the aggregation measures: Theli index, Skewness, Kurtosis, Q3, Q1, Std and Min are not notably different from other measures in association with the **WMC** metric. Similarly from Fig. 10.3b, it can be inferred that Max, Skewness are not notably different from other measures for the **DIT** metric. The two software metrics mentioned here has a different set of aggregation measures which are not making crucial difference to the performance. This depicts the need to consider all the 16 aggregation measures on each of the software metric. Therefore, a total of 16 aggregation measures are applied on each of the 18 source code metrics. This makes the total features considered to be 288 (18 metrics * 16 aggregate measures) which is high dimensional.

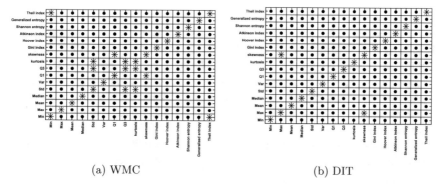

(a) WMC   (b) DIT

Fig. 10.3 Wilcoxon test analysis

10.3.5 Proposed Framework for Feature Selection

All the source code metrics are collected from the java files extracted from the WSDL files. Since these set of source code metrics are used as input, it is very important to remove irrelevant features out of these metrics. Extraction of significant features from the raw set of features is an important step which helps in identifying the set of source code metrics that plays a key role in the detection of the existence of the anti-patterns in the web services. The purpose of the feature selection techniques is to reduce the number of the features and to improve the generalization of models. The best of source code metrics for web service anti-pattern detection are selected by employing the proposed validation framework.

In this work, the best set of source code metrics for the web service anti-pattern prediction are selected using a three step process:

Step-1: Feature Selection with Wilcoxon-signed Rank test (Significant Features (SGF)).

Step-2: Univariate Logistic regression Analysis (Significant Predictors (SGP)).

Step-3: Un-correlated feature analysis using Pearson Correlation Coefficient (Uncorrelated Significant Predictors (UCSGP)).

Wilcoxon Signed Rank Test (SGF): Initially, Wilcoxon Signed Rank Test is applied on the original set of source code metrics. A Wilcoxon signed-rank test is a non-parametric test that can be used to determine whether two dependent samples were selected from populations having the same distribution. It is applied to determine the correlation and influence of the source code metrics on the existence of each type of anti-pattern in a web-service.

Univariate Logistic regression Analysis (SGP): Logistic regression analysis is a statistical analysis method in which ULR analysis has been considered to check the level of significance of each of the object-oriented metric. ULR can be considered as a pre-processing step to an estimator. Therefore, ULR analysis establishes a relation between the independent and the dependent variables. Hence, ULR analysis in anti-pattern prediction investigates whether a source code metric is significant predictor of existing or not-existing anti-pattern in each class. The ULR analysis is applied on the set of source code metrics that are selected by Wilcoxon signed rank test.

Un-correlated Feature Analysis (UCSGP): In this work, Pearson's Correlation Coefficient is used for the cross relation analysis. It is calculated to determine the relationship between different pairs of source code metrics. The Pearson's correlation coefficient, evaluates the strength and direction of linear relationship among two variables. Pearson's Correlation Coefficient is applied on the feature set selected after the ULR analysis.

The proposed feature selection process delivers a unique reduced set of features. Then, the resultant optimal feature set from the considered dataset is tested for accuracy using different classifiers.

10.3.6 Data Balancing Techniques

Class imbalance problem occurs if total number of a class of data (i.e., web services having anti-patterns, in this case) is far less compared to the total number of another class of data (i.e., web services which does not have the anti-patterns). Table 10.1 shows the statistics of five anti-patterns in the dataset contemplated for the experiment. From Table 10.1, it is observed that the dataset appraised is having class imbalance problem. Class imbalance problem may limit the performance of the machine learning and data mining techniques. To overcome this problem, Data balancing approaches are to be used. Balancing can be considered as pre-processing of data which handles the imbalance problem by formulating a balanced training data set, by calibrating the prior distribution for majority and minority classes. In this work two data sampling approaches namely: Random Oversampling approach (RANS) and DownSampling (DWNS) approach are used. The performance of these techniques is evaluated and compared with the original data set (ORG).

Table 10.1 Statistics of anti-patterns (AP) in dataset

	#AP	%AP	#NAP	%NAP
GOWS	21	9.29	205	90.71
FGWS	13	5.75	213	94.25
AWS	24	10.62	202	89.38
CWS	21	9.29	205	90.71
DWS	14	6.19	212	93.81

10.3.7   Classifier Algorithms

In this work, web service anti-pattern detection is considered as a classification problem and 11 classification algorithms such as Logistic Regression (LogR), Decision Tree (DT), Neural Network with three different training algorithms like Gradient Descent (GD), Stochastic Gradient Descent (GDX), Gradient Descent with RBF kernel (GDRBF), Support Vector Machine (SVM) with three different kernels-linear kernel (SVM-LIN), polynomial kernel (SVM-POLY) and RBF kernel (SVM-RBF), Least Square SVM with three different kernels-linear kernel (LSSVM-LIN), polynomial kernel (LSSVM-POLY) and RBF kernel (LSSVM-RBF) and 2 different ensembling techniques such as Majority Voting Ensemble Methods (MVE), Best Training Ensemble (BTE). All this algorithms are used to model different approaches for the detection of anti-patterns.

10.4   Proposed Methodology

Figure 10.4 shows the research framework for developing predictive models for detecting web-service anti-patterns using source code metrics. The dataset downloaded from GitHub has WSDL files from which java files are extracted using a tool called WSDL2JAVA. For each class in Java file, the metrics discussed in Sect. 10.3 are computed using the CKJM tool. The metrics computed are at class level, but the objective of this work is to detect the web service anti-pattern at the system level. To achieve this objective, the metrics are computed at the system level, for which the aggregation measures discussed in Fig. 10.1 are applied on each of the source code metrics computed at the class level. This constitutes the dataset for which the predictive model is built. As shown in Table 10.1, the dataset being considered has data imbalance problem i.e., Class of interest, the web services having anti-patterns is in minority having only 5–10% instances in comparison to 90–95% instances of the majority class which poses technical challenges in building an effective classifier. To deal with this problem, two data sampling techniques i.e., Down Sampling approach (DWSM) and Random Sampling approach (RANSM) are used. After this, feature selection is done using a step-by-step process i.e., first, Wilcoxon signed rank test is applied on the original set of source code metrics (SGF). Secondly, ULR approach is applied on the set of metrics which is the result of first step (SGP). Finally, the Pearson correlation coefficient analysis is applied on the result of the previous step (UCSGP). The 4 sets of source code metrics i.e., SGF, SGP, UCSGP along with the original set of source code metrics (ALF) are considered as input to develop the predictive models for the detection of web service anti-patterns using 11 Machine learning algorithms and 2 ensemble techniques. Finally, the performance evaluation of the developed models are done by using the **Evaluation Metrics**. Although, accuracy and error rate are used as standard evaluation metrics, they are not deemed proper to deal with imbalance classes as the overall accuracy may be biased to the

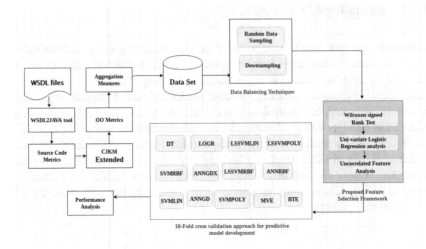

Fig. 10.4 Proposed research framework

majority class nonetheless of the minority class with few samples which lead to poor performance on it. Therefore, AUC (Area Under Curve) is used as evaluation metric, which is popular metric for the imbalanced classes along with F-measure and accuracy.

10.5   Results and Comparative Analysis

10.5.1   Results

The different set of metrics obtained at various steps of proposed framework are used as input to build the models for predicting web service anti-patterns. We are not including the results of the proposed feature selection framework due to the constraint in paper space. These models are trained using 11 classifier learning algorithms and 2 ensemble methods. 10-fold cross validation is used to validate the models developed and performance of these models are computed using two evaluation metrics such as Accuracy and AUC. Table 10.2 details the AUC values obtained for all the classifier techniques applied on the original dataset (All features) and the reduced source code metrics set selected after the application of the proposed frame work for feature selection (UCSGP) as discussed in Sect. 10.3.5. Figures 10.5 and 10.6 shows the boxplots for AUC of the classifier techniques and the various set of source code metrics selected in the proposed feature selection framework for each of the five anti-patterns respectively.

– The AUC values of the predictive models trained using UCSGP is similar to the AUC values of the models built using the original set of source code metrics.
– Figure 10.6 shows that the performance of the models developed using WSRT+ULR+PCC i.e., UCSGP is similar to the performance of the model developed using Allfeatures (ALF) for each of the anti-patterns.
– Figure 10.5 shows that the more than one model that is trained using classifier techniques is showing best results for each of the anti-pattern.

10.5.2 Comparative Analysis

RQ1: What is the capability of various data sampling techniques to predict web service anti-patterns?
In this study to handle the data imbalance problem, two different types of data imbalance techniques such as Downsampling and Random sampling have been considered. The impact and dependability of these techniques are evaluated and compared using Boxplots, Descriptive statistics and statistical test analysis.

Comparison of Different Sampling Techniques Using Descriptive Statistics and Boxplots
Figure 10.7 shows the boxplot for each data sampling techniques for different performance parameters such as AUC and accuracy along with the performance of the model developed using original data. The descriptive statistics for AUC and accuracy for sampling techniques considered is summarized in Table 10.3. From the Fig. 10.7 and Table 10.3, it can be concluded that the random sampling technique achieved better results whereas the model developed with the original data is having the worst performance.

Comparison of Different Sampling Techniques Using Statistical Tests
Random sampling technique (RANSM) is determined as the best sampling technique to handle the data imbalance problem for the data considered, using descriptive statistics and box-plots. After this, Wilcoxon sign rank test is performed to evaluate the statistical significance difference between the pair of different sampling techniques. The null hypothesis of the wilcoxon sign rank test is that, the distinction between the performance of the models developed using different data sampling techniques is not pronounced. The null hypothesis is rejected, if the P-value > 0.05 (denoted by 1) and the null hypothesis is accepted, if the P-value ≤ 0.05 (denoted by 0).

From Table 10.5, it can be seen that considered hypothesis is rejected for all the pairs. From this, it can be concluded that there is a exceptional contradiction between the performance of the anti-pattern prediction models developed using different data sampling techniques.

RQ2: What is the capability of the features selected at various steps of proposed feature selection framework over original features to predict web service anti-patterns?

Table 10.2 AUC of all learning algorithms (overall)

		LOGR	DT	ANNGD	ANNGDX	ANNRBF	SVMLIN	SVMPOLY	SVMRBF	LSSVMLIN	LSSVMPOLY	LSSVMRBF	BTE	MVE
		All features												
ORG	GOWS	0.68	0.81	0.75	0.65	0.69	0.88	0.84	0.50	1.00	0.88	1.00	0.88	1.00
	FGWS	0.67	0.99	0.50	0.71	0.78	0.67	0.98	0.67	1.00	1.00	1.00	0.83	1.00
	DWS	0.99	1.00	1.00	0.93	0.95	1.00	1.00	0.50	1.00	1.00	1.00	1.00	1.00
	CWS	0.89	0.88	1.00	0.84	0.81	0.88	1.00	0.50	1.00	1.00	0.88	0.88	1.00
	AWS	0.83	0.61	0.50	0.51	0.71	0.50	0.98	0.80	1.00	1.00	0.50	0.80	1.00
RANS	GOWS	0.90	0.90	0.90	0.63	1.00	0.80	0.80	0.60	1.00	1.00	1.00	1.00	1.00
	FGWS	1.00	1.00	1.00	0.83	1.00	1.00	1.00	1.00	1.00	1.00	1.00	1.00	1.00
	DWS	1.00	1.00	1.00	0.67	0.50	1.00	1.00	0.67	1.00	1.00	1.00	1.00	1.00
	CWS	1.00	0.90	0.88	0.78	0.75	1.00	1.00	0.63	1.00	1.00	1.00	1.00	1.00
	AWS	0.80	0.88	0.63	0.75	0.70	0.70	0.75	0.70	1.00	1.00	1.00	0.80	1.00
DWNS	GOWS	0.88	0.88	0.75	0.63	0.75	0.90	0.90	0.63	1.00	1.00	1.00	1.00	1.00
	FGWS	1.00	0.83	1.00	0.67	1.00	1.00	1.00	0.67	1.00	1.00	1.00	1.00	1.00
	DWS	1.00	0.83	0.67	0.83	0.83	1.00	1.00	0.83	1.00	1.00	1.00	0.83	1.00
	CWS	0.75	0.88	1.00	0.75	0.63	1.00	1.00	0.60	1.00	1.00	1.00	1.00	1.00
	AWS	0.80	0.65	0.70	0.53	0.80	0.78	0.78	0.75	1.00	1.00	1.00	0.90	1.00
		UCSGP												
ORG	GOWS	0.88	0.88	0.86	0.88	0.73	0.70	0.86	0.75	1.00	0.75	0.88	0.88	1.00
	FGWS	0.67	0.72	0.50	0.78	0.64	0.50	0.74	0.50	0.75	0.83	0.50	0.50	1.00
	DWS	0.83	0.83	1.00	0.97	1.00	0.98	1.00	1.00	1.00	1.00	1.00	1.00	1.00
	CWS	0.84	0.89	0.86	0.84	0.86	0.88	0.99	0.88	1.00	0.90	0.90	0.90	1.00
	AWS	0.60	0.68	0.60	0.60	0.60	0.50	0.50	0.50	0.60	1.00	0.50	0.50	1.00
RANS	GOWS	0.88	0.75	1.00	0.75	0.88	1.00	0.88	1.00	1.00	1.00	1.00	1.00	1.00
	FGWS	1.00	0.83	1.00	1.00	1.00	1.00	1.00	1.00	1.00	1.00	1.00	1.00	1.00
	DWS	1.00	1.00	1.00	1.00	1.00	1.00	1.00	1.00	1.00	1.00	1.00	1.00	1.00
	CWS	0.90	0.90	1.00	0.88	0.88	0.90	1.00	1.00	1.00	1.00	1.00	0.90	1.00
	AWS	0.80	0.70	0.90	0.80	1.00	0.90	0.90	0.88	1.00	1.00	1.00	0.70	1.00
DWNS	GOWS	0.88	1.00	0.90	1.00	0.78	0.90	0.88	1.00	1.00	1.00	1.00	1.00	1.00
	FGWS	1.00	1.00	1.00	1.00	1.00	1.00	1.00	1.00	1.00	1.00	1.00	1.00	1.00
	DWS	1.00	1.00	1.00	1.00	1.00	1.00	1.00	1.00	1.00	1.00	1.00	1.00	1.00
	CWS	0.88	0.90	0.75	0.88	0.88	1.00	0.88	1.00	1.00	1.00	1.00	0.88	1.00
	AWS	0.75	0.78	0.63	0.70	0.70	0.90	0.80	0.70	0.80	0.80	0.80	0.60	1.00

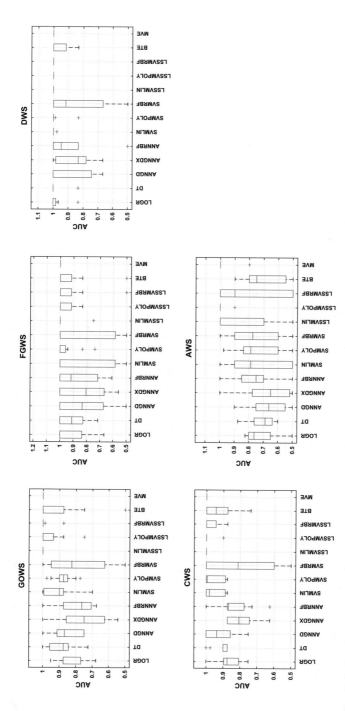

Fig. 10.5 Boxplots for AUC of classifier techniques for each anti-pattern

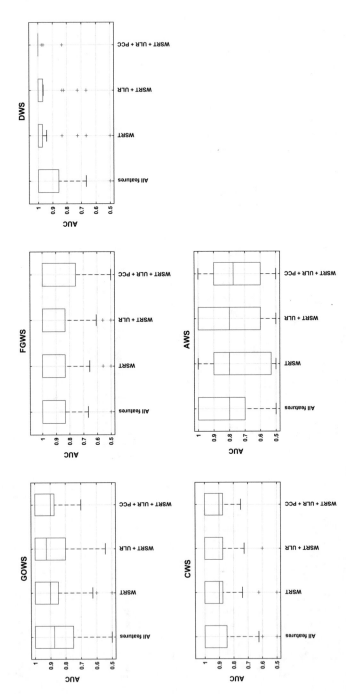

Fig. 10.6 Boxplots for AUC of feature selection techniques for each anti-pattern

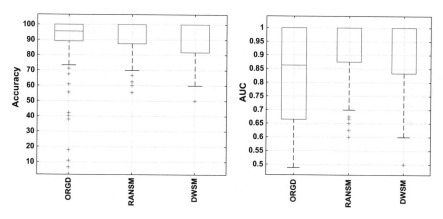

Fig. 10.7 Boxplots for AUC and accuracy of data sampling techniques

Table 10.3 Descriptive statistics of data sampling techniques

	Accuracy						AUC					
	Min	Max	Mean	Median	Q1	Q3	Min	Max	Mean	Median	Q1	Q3
ORGD	6.52	100.00	92.14	95.56	89.13	100.00	0.49	1.00	0.81	0.86	0.66	1.00
RANSM	55.56	100.00	92.92	100.00	87.50	100.00	0.60	1.00	0.93	1.00	0.88	1.00
DWSM	50.00	100.00	91.07	100.00	81.67	100.00	0.50	1.00	0.91	1.00	0.83	1.00

Table 10.4 Descriptive statistics of feature selection techniques

	Accuracy						AUC					
	Min	Max	Mean	Median	Q1	Q3	Min	Max	Mean	Median	Q1	Q3
ALF	10.87	100.00	89.27	95.56	82.50	100.00	0.50	1.00	0.88	0.95	0.78	1.00
SGF	6.52	100.00	91.48	97.83	87.50	100.00	0.50	1.00	0.88	0.92	0.80	1.00
SGP	55.56	100.00	92.89	100.00	87.50	100.00	0.50	1.00	0.89	1.00	0.80	1.00
UCSGP	66.67	100.00	94.54	100.00	90.00	100.00	0.49	1.00	0.89	1.00	0.82	1.00

In this work, the models build using significant features, significant predictors and uncorrelated significant predictors that are selected using wilcoxon sign rank test, uni-variate logistic regression and un-correlated feature analysis are studied. The capability and significance of these techniques are evaluated and quantified using Boxplots, Descriptive statistics and statistical test analysis (Table 10.5).

Comparison of Features Selected from Various Steps of Proposed Feature Selection Framework Using Descriptive Statistics and Box Plots

The descriptive statistics for all the features selected at various steps of proposed feature selection framework for AUC and accuracy is depicted in Table 10.4. Figure 10.8 shows the boxplot for different performance parameters such as AUC and accuracy for the different set of source code metrics selected at various steps. From the Fig. 10.8 and Table 10.4, it can be inferred that the predictive models developed using SGP

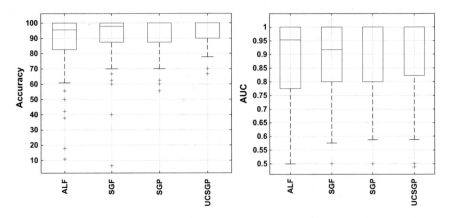

Fig. 10.8 Boxplots for AUC and accuracy of feature selection techniques

Table 10.5 Data sampling techniques: Wilcoxon sign rank

	ORGD	RANSM	DWSM
ORGD	0	1	1
RANSM	1	0	1
DWSM	1	1	0

Table 10.6 Feature selection techniques: Wilcoxon sign rank

	ALF	SGF	SGP	UCSGP
ALF	0	0	0	0
SGF	0	0	0	0
SGP	0	0	0	0
UCSGP	0	0	0	0

and UCSGP are showing a slightly better performance when compared to the models build using SGF and original source code metrics (ALF).

Comparison of Features Selected from Various Steps of Proposed Feature Selection Framework Using Statistical Tests

In this work, Wilcoxon sign rank test is performed to compute the statistical significance difference between the pair of features selected over various steps of proposed feature selection framework. Table 10.6 shows the P-value for the different set of source code metric pairs. The null hypothesis of the wilcoxon sign rank test is that there is no notable dissimilarity between the performance of the models developed using different set of source code metric pairs obtained at various steps of proposed framework. The null hypothesis is rejected, if the P-value > 0.05 (denoted by 1) and the null hypothesis is accepted, if the P-value ≤ 0.05 (denoted by 0).

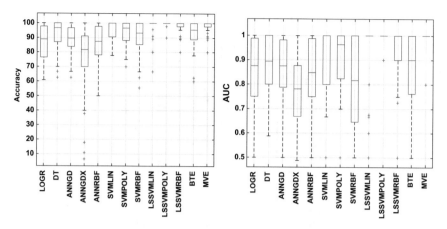

Fig. 10.9 Boxplots for AUC and accuracy of classifier techniques

Table 10.7 Descriptive statistics of classifier techniques

	Accuracy						AUC					
	Min	Max	Mean	Median	Q1	Q3	Min	Max	Mean	Median	Q1	Q3
LOGR	60.87	100.00	86.11	88.89	76.39	97.78	0.50	1.00	0.84	0.88	0.75	0.99
DT	62.50	100.00	91.27	96.71	87.08	100.00	0.59	1.00	0.88	0.89	0.80	1.00
ANNGD	62.50	100.00	89.63	89.57	83.89	96.71	0.50	1.00	0.84	0.88	0.79	0.98
ANNGDX	6.52	100.00	77.64	81.67	70.00	91.11	0.49	1.00	0.78	0.78	0.67	0.88
ANNRBF	50.00	100.00	86.77	87.50	78.02	97.78	0.50	1.00	0.83	0.85	0.75	0.99
SVMLIN	77.78	100.00	95.21	100.00	90.56	100.00	0.50	1.00	0.89	1.00	0.80	1.00
SVMPOLY	70.00	100.00	92.98	96.69	88.19	100.00	0.50	1.00	0.90	0.96	0.82	1.00
SVMRBF	55.56	100.00	90.34	93.33	85.42	100.00	0.50	1.00	0.80	0.82	0.65	1.00
LSSVMLIN	66.67	100.00	98.53	100.00	100.00	100.00	0.50	1.00	0.96	1.00	1.00	1.00
LSSVMPOLY	90.00	100.00	99.26	100.00	100.00	100.00	0.90	1.00	0.99	1.00	1.00	1.00
LSSVMRBF	80.00	100.00	97.62	100.00	97.78	100.00	0.50	1.00	0.92	1.00	0.90	1.00
BTE	60.00	100.00	93.61	95.56	88.89	100.00	0.50	1.00	0.87	0.90	0.76	1.00
MVE	80.00	100.00	97.59	100.00	97.78	100.00	0.80	1.00	1.00	1.00	1.00	1.00

From Table 10.6, it can be seen that null hypothesis is accepted for all the pairs considered as the comparison points. Hence, it can be concluded that the performance of the prediction models developed using different set of source code metric pairs obtained at various steps of proposed framework are similar. This means that the models developed using different set of features obtained from various steps have the same performance accuracy. Hence, the feature selection technique which takes the least number of inputs should be considered for building the predictive model.

RQ3: What is the impact of various classifier techniques to predict web service anti-patterns?

In this work, eleven classifier techniques along with two ensemble techniques namely Majority Voting Ensemble (MVE) and Best Training Ensemble (BTE) have been

Table 10.8 Classifier techniques: Wilcoxon sign rank

	LOGR	DT	ANNGD	ANNGDX	ANNRBF	SVMLIN	SVMPOLY	SVMRBF	LSSVMLIN	LSSVMPOLY	LSSVMRBF	BTE	MVE
LOGR	0	0	0	1	0	1	1	0	1	1	1	0	1
DT	0	0	0	1	0	0	0	1	1	1	1	0	1
ANNGD	0	0	0	1	0	1	1	0	1	1	1	1	1
ANNGDX	1	1	1	0	1	1	1	0	1	1	1	1	1
ANNRBF	0	0	0	1	0	1	1	0	1	1	1	1	1
SVMLIN	1	0	1	1	1	0	0	1	1	1	0	0	1
SVMPOLY	1	0	1	1	1	0	0	1	1	1	1	0	1
SVMRBF	0	1	0	0	0	1	1	0	1	1	1	1	1
LSSVMLIN	1	1	1	1	1	1	1	1	0	0	1	1	1
LSSVMPOLY	1	1	1	1	1	1	1	1	0	0	1	1	0
LSSVMRBF	1	1	1	1	1	0	1	1	1	1	0	1	1
BTE	0	0	1	1	1	0	0	1	1	1	1	0	1
MVE	1	1	1	1	1	1	1	1	1	0	1	1	0

considered to train the models for predicting web services. The capability and significance of these techniques are evaluated and quantified using Boxplots, Descriptive statistics and statistical test analysis.

Comparison of Classification Techniques Using Descriptive Statistics and Box Plots
The descriptive statistics for all the classifier techniques along with ensemble techniques for AUC and accuracy is depicted in Table 10.7. Figure 10.9 shows the boxplot for different performance parameters such as AUC and accuracy for all the classifier techniques. From the Fig. 10.9 and Table 10.7, it can be concluded that the Majority Voting Ensemble (MVE) achieved the best performance when compared to all the other techniques, whereas ANNGDX performed poorly.

Comparison of Classification Techniques Using Statistical Tests
In this work, Wilcoxon sign rank test is performed to compute the statistical significance difference between the pair of different classifier. Table 10.8 shows that the P-value for all the classifier technique pairs. The null hypothesis of the wilcoxon sign rank test is that the performance of the models developed using different classifier techniques is not impressively divergent. The null hypothesis is rejected, if the P-value > 0.05 (denoted by 1) and the null hypothesis is accepted, if the P-value ≤ 0.05 (denoted by 0). From Table 10.8, it can be seen that null hypothesis is rejected for most of the pairs considered as the comparison points. Hence, it can be concluded that the performance of the anti-pattern prediction model developed using different classifier techniques are contrasting.

10.6 Conclusion

The goal of the research is to find the impact of the source code metrics which defines the internal structure of the software on the prediction of web service anti-patterns. In this work, we empirically computed, analyzed and compared the performance of 13 classifier techniques and 2 data sampling techniques using the various set of source code metrics selected at different stages of proposed feature selection framework for the detection of anti-patterns in web services. The main findings of this study are summarized below:

– Random sampling technique provides relatively better performance in predicting web service anti-patterns when compared to the models built using down sampling technique.
– Majority Voting Ensemble technique provides better performance in detecting anti-patterns compared to the models built by other classifiers.
– Selected un-correlated significant predictors provides relatively similar performance for predicting web service anti-patterns to the model built using all the source code metrics.

– The web service anti-pattern prediction models developed using different data
  sampling techniques and classifiers are remarkably different, although the models
  developed using different set of features i.e., SGF, SGP, UCSGP and ALF is same.

References

1. Král, J., Zemlicka, M.: Crucial service-oriented antipatterns, vol. 2, pp. 160–171. International
   Academy, Research and Industry Association (IARIA) (2008)
2. Brown, W.H., Malveau, R.C., McCormick, H.W., Mowbray, T.J.: AntiPatterns: Refactoring
   Software, Architectures, and Projects in Crisis. Wiley, Hoboken (1998)
3. Dudney, B., Asbury, S., Krozak, J.K., Wittkopf, K.: J2EE Antipatterns. Wiley, Hoboken (2003)
4. Rodriguez, J.M., Crasso, M., Zunino, A., Campo, M.: Automatically detecting opportunities
   for web service descriptions improvement. In: Conference on e-Business, e-Services and e-
   Society, pp. 139–150. Springer (2010)
5. Moha, N., Palma, F., Nayrolles, M., Conseil, B.J., Guéhéneuc, Y.-G., Baudry, B., Jézéquel, J.-
   M.: Specification and detection of SOA antipatterns. In: International Conference on Service-
   Oriented Computing, pp. 1–16. Springer (2012)
6. Ouni, A., Gaikovina Kula, R., Kessentini, M., Inoue, K.: Web service antipatterns detection
   using genetic programming. In: Proceedings of the 2015 Annual Conference on Genetic and
   Evolutionary Computation, pp. 1351–1358. ACM (2015)
7. Palma, F., Moha, N., Tremblay, G., Guéhéneuc, Y.-G.: Specification and detection of SOA
   antipatterns in web services. In: European Conference on Software Architecture, pp. 58–73.
   Springer (2014)
8. Nayrolles, M., Palma, F., Moha, N., Guéhéneuc, Y.-G.: Soda: a tool support for the detection of
   SOA antipatterns. In: International Conference on Service-Oriented Computing, pp. 451–455.
   Springer (2012)
9. Marinescu, R.: Detection strategies: metrics-based rules for detecting design flaws. In: 20th
   IEEE International Conference on Software Maintenance, 2004. Proceedings, pp. 350–359.
   IEEE (2004)
10. Chidamber, S.R., Kemerer, C.F.: A metrics suite for object oriented design. IEEE Trans. Softw.
    Eng. **20**(6), 476–493 (1994)
11. Vasilescu, B., Serebrenik, A., van den Brand, M.: You can (2011)

Printed in the United States
by Baker & Taylor Publisher Services